SUSTAINABLE AQUACULTURE

SUSTAINABLE AQUACULTURE

Edited by

John E. Bardach
The East-West Center
Honolulu, Hawaii

JOHN WILEY & SONS, INC.

New York • Chichester • Weinheim • Brisbane • Singapore • Toronto

This text is printed on acid-free paper.

Library of Congress Cataloging in Publication Data:

Sustainable aquaculture / edited by John E. Bardach.
p. cm.
Includes bibliographical references and index.
ISBN 0-471-14829-6 (cloth : alk. paper)
1. Aquaculture—Environmental aspects. 2. Aquaculture—Management. I. Bardach, John E.
SH135.S87 1997
639′.8—dc20 96-44727

Printed in the United States of America
10 9 8 7 6 5 4 3 2 1

CONTENTS

PREFACE

By dealing with many aspects of aquaculture, this book broaches two important topics of human survival—food supplies and sustainability; therefore, it ought to be of interest to a broad public. Although not primarily a technical book, it contains some technical information together with discussions of a social and economic nature. It examines the questions, Can aquatic animal husbandry keep the promise many hold for it of closing the gap between the increasingly larger shortfalls of fish caught from the sea and the increasing need for high-grade animal protein worldwide? It must be stressed that aquaculture is developing very rapidly, especially in Asia. Therefore, this book must be looked at as a landmark or way station rather than as a report on the state of the art.

The need for such a book became evident in a symposium with the same title held by the Pacific Congress of Marine Sciences and Technology (PACON), in Honolulu in June 1995. This book is not a proceeding of this meeting, but it became clear that a written treatment of the many facets of human endeavor that make aquaculture possible would be useful. I therefore asked several colleagues, some of whom had contributed to the PACON meeting, to help me put together such a compendium; what follows in the remainder of the Preface is an overview of the resulting volume.

Since it is not only persons in the aquaculture industry but also economists, administrators, and politicians who help to develop aquatic farming and husbandry, the text begins in Chapter 1 with the framework of fish as food, where of course "fish" encompasses shrimp, oysters, and other edible invertebrates. This chapter also gives an overview of the various techniques for rearing aquatic animals under fully or partially controlled conditions. Finfish are the main subject because they are the most important by volume, in spite of cultured shrimp having now caught media attention. Algae are treated only cursorily as a condiment food, and treated as such in only a part of the world. Their use in industry is excluded.

Since Asia is the most populous continent and also the one blessed with much freshwater runoff (the Amazon notwithstanding), it is not surprising that it is first in fish production, including aquaculture, and that much of it comes from fishponds, lakes, and rivers rather than from the oceans or the coastal regions. Meryl Williams, Director General of the International Center for Living Aquatic Resources Management (ICLARM), writes in Chapter 2 about aquaculture and sustainable food security in the developing world. Since ICLARM is a member of the Consultative Group for International Agricultural Research (CGIAR), her chapter contains many cross-references to land-animal husbandry. Surveying aquaculture continent by continent, she points to social and economic obstacles such as lack of capital, education, infrastructure, and the like, as well as to increasing limitations for aquaculture in coastal regions because of population pressure with attending urbanization and its water demands. Nevertheless, food will still have to come from nonurban regions in spite of the fact that more than half the world's people will live in cities, not too far into the next century.

The ocean beyond nearshore waters has long beckoned as a next area to be used for cultivation. Robert Stickney, Director of the Sea Grant Program at Texas A&M University, deals in Chapter 3 with the challenge of extending aquaculture beyond the shallow parts of the coastal zone where it is now practiced. One cannot be too optimistic about these opportunities on grounds of technical and operational constraints; winds and waves are inimical and corrosion destroys materials, even though obtaining entitlement to an offshore culture site may be easier than inshore. Even so, these legal matters have only the precedent of oil rigs, which may offer opportunities for attachment of cages and comparable rigs, especially when they do not produce oil anymore. All such sites suffer from difficulty of access, and access may have to be more frequent for aquacultural than for oil installations. Since shelled mollusks and crustaceans are tied to the bottom, only finfish are left for the offshore culture schemes. Considering that most desired marine fish are carnivores, the keeping of which in high concentrations requires the provision of feed, we certainly arrive at limitations to a "blue food revolution." Technological optimists will have a hard job finding fault with Dr. Stickney's realistic treatment.

Water pollution is a worldwide problem that affects all human activities, aquaculture being only one of them. But keeping many animals in one space influences their growing conditions and also in some cases the environment beyond their confinement. I write in Chapter 4 on pollution and biodiversity, dealing mostly with self-pollution and influences on the aquatic environment close to aquaculture facilities. Pollution by aquaculture is still only sporadic, but upcoming intensification and population increases threaten to make it more prevalent. Also, on the coast, aquacultural installations have less protection from pollution than they do inland, where ponds are the main bases of production. Biodiversity is considered in the same chapter as pollution because its reduction is one byproduct of biological pollution. Aquaculture can play an important role here, through escape of cultivars into the surrounding water, disease transmission, and (albeit very locally) through preempting space and altering environmental conditions in the water. Some aspects of biodiversity are also treated in Chapter 5, on biotechnology.

Chapter 5, by Edward M. Donaldson of the Vancouver Laboratory of Canada's Ministry of Fisheries and Oceans, which has long done applied research on biotechnology, considers the role of this branch of applied science in sustainable aquaculture. It is the most technical chapter in the book and also looks quite far into the future. Biological chemistry, physiology, and genetics, skillfully blended and selectively applied, could convert an often hazardous trade (which aquaculture still largely is) into a quasi-engineered, safe production of quality-controlled aquatic food commodities. This is clear from perusing Donaldson's discussion of a subject matter that sometimes may appear to have a touch of science fiction. In fact, sustainability will rest on applied science and engineering joined to the social skills on which wise husbandry of natural resources increasingly depends. Land-animal husbandry, especially poultry rearing, has already profited from biotechnology, and we can now rely on factory-like production of chickens or pigs; biotechnology, so Donaldson argues, may be more easily applied to aquatic organisms that have eggs and early stages more accessible to manipulation than to large, shelled eggs of birds or embryos *in utero*. The chapter also stands out because of its exhaustive bibliography.

The chapters on economic decision making by Yung Shang and Clem Tisdell (Chapter 6) and on modeling in planning for aquaculture by PingSun Leung and Omar El Gayar (Chapter 7) are by agricultural economists, three at the University of Hawaii, and one at the University of Queensland in Australia, are best considered together. Aquaculture must make a profit (which even subsistence enterprises hope to do after a fashion) and maintain its base, which includes water quality and in some cases broodstock integrity. True, much from the economics of agriculture can be applied to aquaculture, but special characteristics of the fluid watery medium introduce the need to give attention to very particular externalities. Increasing conservation needs and multiple pressures on natural resources make it necessary to eliminate externalities by making them part of the operation itself (e.g., the reusing of effluents). This is a challenge to economic planning, which is treated from both micro and macro viewpoints, as well as to technology and management where modeling can furnish some shortcuts. Both chapters touch on these problems albeit from somewhat different vantage points.

Pressure on the various resources on which aquaculture depends is a prominent theme in the various chapters; it is not surprising that Chua Thia-Eng, of the International Maritime Organization, writes in Chapter 8 of integrated coastal management (ICM) as the logical tool to accommodate various uses. It is "easier said than done," though, to create a mechanism for fair representation of the various interests, especially since they may have very different time horizons prominent in their expectations. And this is especially true for small farmers against more capitalized and influential occupants. Dr. Eng stresses that integration of coastal stakeholders must have a local base and works best from the bottom up—all stretches of coast are, after all, different from one another in their characteristics—but that it helps if the governmental unit involved facilitates the creation of an administrative structure to make management not only ecologically sound but also fair and equitable. Under special circumstances parts of a coast might be set aside for aquaculture if only as relatively small parks and the like, though this has hardly been done successfully anywhere.

Accommodating new installations in about-to-be-developed coastal areas is easier than dealing with existing ones. These latter have had the tendency, in Asia and elsewhere, to exceed the carrying capacities for aquaculture. The chapter describes a (few) relatively successful, and in contrast not so successful, examples of ICM. Inasmuch as the concept is a relative "newcomer" and officials might have to be prevailed on to give up power or to share it, it is easier to plan for ICM than to bring it into successful operation. The latter may take many years, if not decades, but not to try to do so is likely to have long-term adverse consequences.

John Corbin and Leonard Young, of Hawaii's Aquaculture Development Program, which is considered to be relatively forward-looking among such efforts in the United States, take an enlightened administrative attitude to their mission. In Chapter 9, they start with site assessment and planning at the micro and macro levels and then examine hierarchies of bureaucracies (permits, enforcement, and so on) necessary to enable aquaculture to function in an economy of resource utilization. Corbin and Young deal with matters in advanced and in developing countries; an enterprise has to function as a business, whether small or large, even while satisfying the goals of sustainability. They also write about government assistance, finances, and incentives. This chapter and Chapter 8 are useful companion pieces that complement one another.

Finally, in the postscript I attempt a prognosis in light of the various other chapters. The question has to be asked: How optimistic can we be that aquaculture will fulfill its promise? In this final discussion I deal with the interplay of the biological and technical bases for the rearing of aquatic organisms with the many social and economic forces that impinge on the domestication of these organisms—for that is what is at stake. A comparison of aquaculture with avian and mammalian animal husbandry is certainly warranted. However, we must never forget that life in the water is very different from life on land, that we can control that medium and its denizens only to a certain extent, and that there are quite naturally a great many more candidates for aquatic rearing than the few domesticated terrestrial creatures that currently give us our protein food supply. Still, I am quite optimistic: in my youth chicken was a Sunday luxury fare; it has since become a relatively cheap and fairly universally available food item. The same trend could apply to several species complexes in the next several decades with luck, perseverance, and attention to preserving our life-support bases.

This book may be used in several ways: specialists might read thoroughly only those chapters that pertain to their areas of expertise and browse through the others. Managers and decision makers, in turn, might also be selective, focusing their emphasis on economics and various planning aspects, such as coastal-zone planning or administration and regulation. The whole book is recommended for educators and students with interests in natural resources or food and water problems. Some repetition, albeit highlighted by the different fields of specialization of the writers, is unavoidable in the organization chosen for the book, but this organization gives the book a special flavor; its horizontal integration makes it suitable for a wide audience.

I am grateful to my cooperative chapter authors for their contributions. I also want to thank the Program on Environment and the Office of Research and Education of the East-West Center, and especially Jeni Miyasaki, for making it possible for me to work on this book.

JOHN E. BARDACH, EDITOR

East-West Center
Honolulu, HI

CONTRIBUTORS

JOHN E. BARDACH, East-West Center, 1777 East-West Road, Honolulu, HI 96848, fax: (808) 944-7502

JOHN S. CORBIN, Aquaculture Development Program, Department of Land and Natural Resources, State of Hawaii, 335 Merchant Street, Room 348, Honolulu, HI 96813, fax: (808) 587-0033

EDWARD M. DONALDSON, Vancouver Laboratory, Department of Fisheries and Oceans, 4160 Marine Drive, West Vancouver, BC V7V 1N6, Canada, fax: (604) 666-3497

OMAR F. EL-GAYAR, IAAEM, Department of Agricultural & Resource Economics, University of Hawaii, 3050 Maile Way, Gilmore Hall 115, Honolulu, HI 96822, fax: (808) 956-2811

PINGSUN LEUNG, IAAEM, Department of Agricultural & Resource Economics, University of Hawaii, 3050 Maile Way, Gilmore Hall 115, Honolulu, HI 96822, fax: (808) 956-2811

YUNG C. SHANG, Department of Agricultural & Resource Economics, University of Hawaii, Gilmore Hall 124, Honolulu, HI 96822, fax: (808) 956-2811

ROBERT R. STICKNEY, Texas Sea Grant College Program, Texas A&M University, 1716 Briarcrest, Suite 702, Bryan, TX 77802, fax: (409) 845-7525

CHUA THIA-ENG, U.N. Development Programme/International Maritime Organization, P.O. Box 2502, Quezon City, Metro Manila 1165, Philippines, fax: (63-2) 635-4216; 635-5843

CLEM A. TISDELL, Department of Economics, The University of Queensland, Australia, fax: (61-7) 365-6666

MERYL J. WILLIAMS, International Center for Living Aquatic Resources Management (ICLARM), NCPO Box 2631, Makati, Metro Manila 0718, Philippines, fax: (63-2) 812-3798

LEONARD G. L. YOUNG, Aquaculture Development Program, Department of Land and Natural Resources, State of Hawaii, 335 Merchant Street, Room 348, Honolulu, HI 96813, fax: (808) 587-0033

1 Fish as Food and the Case for Aquaculture

JOHN E. BARDACH

INTRODUCTION

Throughout the many-millennia-long hunting-and-gathering period of human prehistory, fish and shellfish contributed prominently to the food supplies. With the evolution of agriculture and animal husbandry, the relative importance of both fish and game declined, but animal protein sources of aquatic origin still remain crucial to human welfare in many parts of the world. Only a small percentage of the world's population—about 6%—are vegetarians by choice; they include adherents to the Hindu and Jain religions and some others who manage to compose salutary diets without animal protein. The rest of humanity obtains certain of the essential amino acids from the flesh of domesticated animals and to a lesser degree from fish. On an overall global scale, fish and fishery products make up about 16% of animal protein intake. In developing countries, where meat is mostly scarce, the percentage is, not unexpectedly, higher—about 20%, with Latin America at 9%, Africa at 19%, and Asia at 29% (Anonymous, 1995). The more affluent, developed countries have a lower intake of fishery products—about 14% (FAO, 1992). These summary figures cover up many geographic, economic, and cultural differences. For instance, both Mongolia and the Maldives have higher animal protein intakes than their regional averages, but in the case of the former the intake is based on livestock rearing and in the case of the latter it derives from a historic fishing tradition. By the same token, beef is the meat fare of the Argentineans, while the Japanese obtain more than half of their presently copious animal protein diet from their fisheries (FAO, 1992). Some other Asian nations also have a high component of fish in their diets; however, the basic amounts that are available differ widely in parts of the same country, especially the largest countries such as China and India.

In this century the world fish catch grew from a very rough estimate of 4 million metric tons (MMT) in 1900 (Borgstrom, 1962) to nearly 100 million (FAO, 1995), or about 20-fold, by 1995, while the world's people increased from about 1.6 to nearly 6 billion, or about a factor of 4. The fish and shellfish are essentially nature's

Sustainable Aquaculture, Edited by John E. Bardach
ISBN 0-471-14829-6

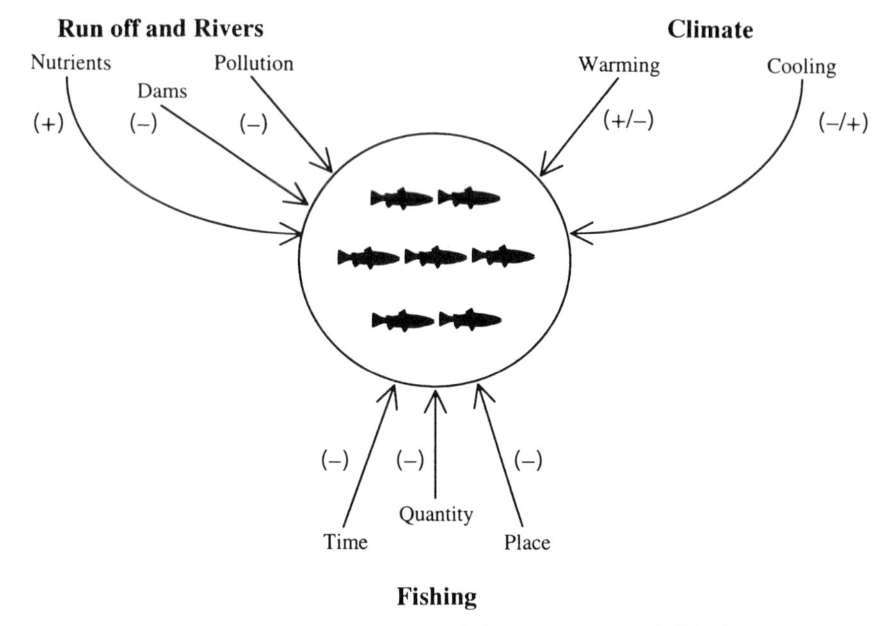

Figure 1.1. Human and natural influences on coastal fisheries.

bounty—that is, they are neither husbanded nor tended—but they are subject to both climatic and human influences that are mostly stressful (Figure 1.1); it is not surprising that we should have reached the extent of their natural limits. And the future demand of humankind for protein from the waters of the globe will have to come increasingly from husbandry or culture practices.

This is well illustrated if we look at the oceans not just as a large expanse of water that covers 70% of the planet's surface, but as various different ecosystems each with its own characteristics of biological production. In fact, over 90% of the marine fish catch comes from reaches that are less than 200 miles from the shore (FAO, 1981). In the vast open ocean, tuna and billfishes, though high-priced catches, make up less than 5% of the total fish production. These two kinds of fish chase their food, which is mostly fish that occur in scattered patches, though the base of their food chain, planktonic algae, is widespread but diffuse. It requires four to five or more steps up the food chain—via zooplankton, small invertebrates, fishes, and larger prey—to sustain tuna. As only about 10% of material and energy is transferred in each step, harvesting at the top of this food chain cannot be done economically now, nor is it likely in the future.

It is estimated that 69% of ocean fish stocks and stock aggregates are fished to capacity or overfished (FAO, 1995); these come from nearshore upwelling systems and tropical or temperate shelves and banks, including coral reefs. In these shallow ecosystems, between 25% and 35% of primary algae production is represented in the catch (Pauly and Christensen, 1995). This figure is (alarmingly) close to that for the land areas of the globe, where an estimated 35% to 40% of primary production (plants) is now used directly and indirectly (Vitousek et al., 1986). There remain, then, lakes and rivers that make up less than 1% of the global surface, as opposed to 70% for the oceans; the freshwaters furnish about 7% of our untended fish harvest. Considering the multiple pressures on them, a natural increase here is not likely either.

The shortfall in protein of aquatic provenance will thus be considerable in the next century, and aquacultural food production would have to be greatly increased. Agricultural, industrial, and urban competition notwithstanding, much of this development will be in fresh water, where aquaculture already has a successful history, but it is clear that new technologies will also enable further forays into controlled production of fish and shellfish in the sea.

Some points made in this overview are underscored by an examination of the last 10 years of global fish production (FAO, 1995) by the several of the most important fishing nations. Perhaps most notable is the rise of China into first position, mainly due to the rapid growth of fresh- and brackish-water aquaculture in the deltas of the Pearl, Yangtse, and Yellow Rivers. Especially in the Yangtse plain with its many lakes and canals, the state fostered integrated land and water management and polyculture—the rearing together of suitable fish species (Figure 1.2), as well as avian and mammalian stock. A change from communal farms into cooperatives and an incipient profit orientation were also important, as was a rise in the culture of scallops, mussels, oysters, and shrimp along China's coastlines.

The former Soviet Union, which long vied with Japan for first position in fisheries, in 1993 fell back to fifth place. Internal economic difficulties did not permit it to continue the earlier wide-ranging efforts of its fishing fleets, which occurred not only in the North Pacific but also as regular trips with factory ships within the 200-mile economic zones, especially of African nations; now even the fleets could not be kept up as they were earlier. Peru and Chile, which have long been important fishing nations, kept the leading positions afforded to them by bordering the world's most important upwelling system, which produces myriads of plankton-feeding anchovies and sardines, the basis of the largest portion of the world's fish-meal production. Pilchards and jack mackerel, which can also peak in their waters, also increased in the catches. Severe fluctuations in the plankton, and on the anchovy catch immediately dependent on it, occur when the cold, fertile Peru Current is displaced by warmer, nutrient-poor surface water (El Niño), an event in turn related to wider variations in the Pacific-wide current patterns and to overall global climatic perturbances. Other fisheries are also relying increasingly on low-value species, which are by and large low on the food chain where short-term climatic fluctuations are quickly felt (Chikuni, 1987). It is obviously important to be able to predict these changes, as they may occur within several-year intervals, and fishing fleets and shore installations are intended to operate on longer time scales. Prediction of an El Niño, now possible with a horizon of several months to a year, permits anticipatory management, at least to some extent.

India, South Korea, Indonesia, Thailand, and Vietnam are rising and populous nations that during the 20th century have steadily increased their fish production;

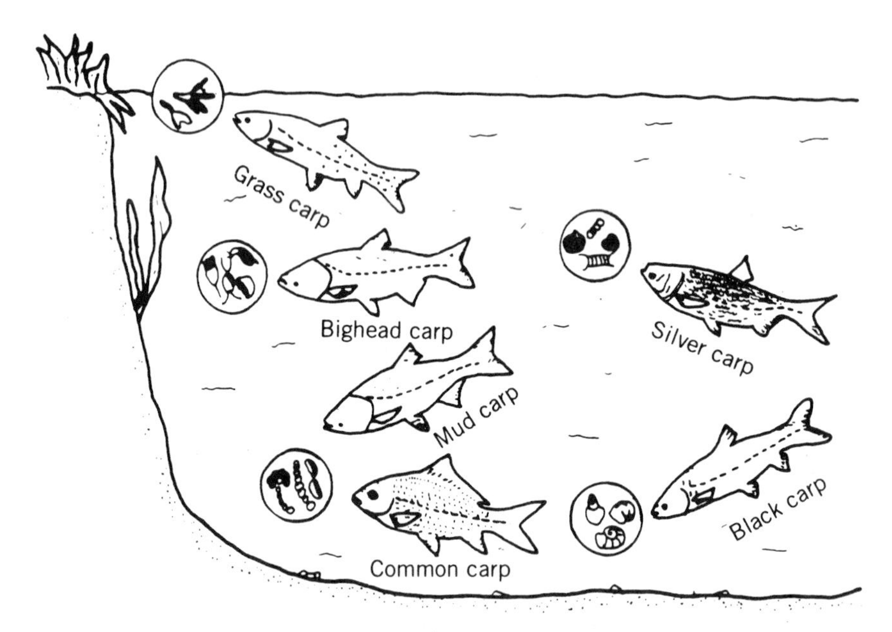

Figure 1.2. Habitat and feeding niches of the principal species in classical Chinese carp culture. Grass carp (*Ctenopharyngodon idella*) feed on vegetable tops. Bighead carp (*Aristichtys nobilis*) feed on zooplankton in midwater. Silver carp (*Hypophthalmichtys molitrix*) feed on phytoplankton in midwater. Mud carp (*Cirrhinus molitorella*) feed on benthic animals and detritus, including grass carp feces. Common carp (*Cyprinus carpio*) feed on benthic animals and detritus, including grass carp feces. Black carp (*Mylopharyngodon piceus*) feed on mollusks. (Adapted from J. E. Bardach, T. H. Ryther, and W. O. McLarney, Aquaculture: Farming and Husbandry of Fresh Water and Marine Organisms (1972), John Wiley & Sons, Inc., New York).

this is due to development of and investment in vessel and capturing technology that has permitted them to exploit stocks that were more lightly fished in the past. These countries fish mostly on shelves, banks, and seamounts, where limits will undoubtedly be reached in the not too distant future. Denmark, Iceland, and Spain, all nations with long fishing traditions, mainly in the North Atlantic but also worldwide, show relatively steady and sometimes, like Norway, declining and rising fish tonnages.

Yet, considering the parlous state of many of the world's fisheries, it is interesting that the management of fish stocks has long been related to the concept of sustainability. Maximum sustainable yield (MSY) is based on the notion that each species or stock has its surplus that can be taken without impairing its reproductive potential, somewhat like interest on invested capital (Gulland, 1971). This biologically derived concept has influenced the setting of catch quotas, but it is only one side of a multifactoral equation on which management is based. The other checks on fish populations besides fishing are social, economic, cultural, and political, aside from the climatic ones that cannot yet be assessed satisfactorily. Thus the generally used controls to curb excessive fishing, such as limiting season, gear, and area of access to what is in essence a common property resource with weak guardians and avid takers, have all too often come to naught. But even with controls that would permit the recovery of many fisheries under sustainable common-property regimes, limits in the natural world suggest greater reliance on culture will be necessary to meet future demand. There the common-property constraints of too many takers on a naturally limited resource are replaced by the need to control the means of production rather than access to the bounty of nature.

SUSTAINABILITY AND PATTERNS OF AQUACULTURE

The World Commission on Environment and Development (WCED) was established by the United Nations in 1983 of eminent persons from politics, business, and academe, and is also known as the Brundtland Commission after the name of its chairperson. The WCED has advocated aquaculture as one of the measures that would help attain sustainable development, (WCED, 1987). In the many discussions since the WCED met about what sustainability really means it has become clear that the term can have many meanings depending on whether one sees the world through ecological or economic glasses. As long as populations grow and economic conditions improve for many, the most sustainable development will be one that attains the best possible relationship of the forces active in local and regional dynamic cultural and economic systems as well as in larger dynamic, but normally slower-changing, ecological systems. To be sustainable these systems must allow (a) human life to continue indefinitely, (b) human individuals to flourish, and (c) human cultures to develop; at the same time the effects of human activities must remain within bounds so as not to destroy the diversity, complexity, and function of the ecological life support system (Costanza, 1991).

The FAO (1991) has defined sustainable development in a similar manner, meaning it to be

> the management and conservation of the natural resource base, and the orientation of technological and institutional change in such a manner as to ensure the attainment and continued satisfaction of human needs for present and future generations. Such sustainable development conserves land, water, plant and animal genetic resources, is environmentally non-degrading, technically appropriate, economically viable and socially acceptable.

The social acceptability component has often been overlooked; aquaculture development requires an appropriate range of institutional facilities and therefore investment and support for them. It also requires consideration of cultural forces that may assist or hinder its progress. If these are disregarded, the practice of rearing organisms will have difficulties reaching its potential even with the best intent of the individuals who may want to engage in it (Ruddle, 1993).

In fact, in a comparative study done for the U.S. Congress' Office of Technology Assessment on commonalities in national aquacultural successes and failures, the crucial role governments play stands out in every case of success (Katz, 1995). The aquaculture industry is healthy where all or most economic factors, legal policies (legislative, judicial, and enforcement), government/private-sector cooperation, import-export policy, and customs practices concerning it are in place. If governments recognize the role the environment now plays in development, there is additional positive influence on aquaculture.

These attributes of sustainability suggest well-planned changes in the management of natural resources; in animal production on land and in the water, it means using less energy, recycling (Edwards, 1993), reduction and substitution of raw-material inputs, and the like. This is well exemplified by the practice in China of growing compatible fish species together (Figure 1.2). High returns, as percentages of variable costs, have been achieved in such schemes, adding ducks (Jhingran and Sharma, 1980); knowledge of ecological relations is, however, necessary, and the input of labor is not low (see also Chapter 6). The present emphasis on maximizing the efficiency of producing a single food commodity, as is the practice in many economies, has also given us very high harvests per unit surface area. North American wheat, catfish, cotton, trout, and shrimp are some examples. The environmental costs of such monocultures, though, are well known even if climatic forces may exacerbate them, as they did in the Dust Bowl years of the American Plains.

The quest for environmental sustainability implies a move away from monocultures in a viable transition to acceptable, complex, multigoal-oriented bioproduction systems. Wherever these can be instituted, such changes would mean a shift in the values attributed to the systems' components even though certain large monocultures would certainly remain. Such a shift is not likely to happen soon, but there are instances of successful, large-scale reuse of nutrients, both in agriculture and aquaculture, and the integration of agricultural and aquacultural systems has been shown to be technically feasible (Brune, 1993).

This book will stress what this may mean for aquaculture—in essence to assist in our understanding of the changes necessary to allow aquaculture to contribute to economic betterment in a healthy environment. To do this, we must first describe several basic attributes of aquaculture, defining it as forms of intervention in the rearing process of aquatic animals and plants to enhance production. Included here are breeding, stocking, feeding, and protection from predators as well as assuring the health of the cultivars. Aquaculture also implies individual or corporate ownership, as opposed to fisheries, which harvest common-property resources (FAO, 1995). The commodities of aquaculture compete directly with those of fisheries in the food marketplace, while the processes of their production are more akin to those of agriculture, paramount among these being, in most instances, ownership of stock and control of the farming sites. This contrast raises the question of where national oversight over aquaculture development should be placed—in agriculture or fisheries departments, as is now the case in most nations? The matter will be treated later in the book.

We will deal here only with the farming of aquatic animals (fishes, mollusks, and

TABLE 1.1. World Aquaculture Production 1993: Tonnage and Various Percentages

	Fish	Mollusks	Crustaceans	Plants (Mostly Algae)
Million metric tons (MMT)	11.2	4.1	0.93	5.3
% Weight (of total)	49.5	18.2	4.1	27.7
% Value (of total)	52.8	15.0	12.5	16.5
% Wt. fresh water	96.5	Negligible	6.8	—
% Value fresh water	85.5	Negligible	5.0	—

Total tonnage of animals was 16.23 MMT, of which 85.5% came from Asia.
Source: FAO (1995).

crustaceans) reared for food in fresh, brackish, or marine waters; the last case is sometimes called mariculture. Algae for food and production of industrial colloids should at least be mentioned because they are an important maricultural crop, especially in Asia as Table 1.1 (an overview of world aquaculture production) illustrates. Well over 90% of crustaceans in volume are reared in brackish and marine waters, while the tonnage of fishes of aquaculture comes to over 95% from freshwater environments, where control is easier and adverse forces such as weather and corrosion are less severe. Virtually no mollusks of commerce, cultured or not, come from fresh water. (FAO, 1995).

Aquatic animal husbandry may be undertaken with full control over all phases of a life cycle, which include breeding, seed or larval rearing, and feeding the various stages from fry to fingerlings to adults. Also included are preventing their escape, protecting them against diseases and predators, and attending to the quality of their water. Thus we may describe full-culture operations as "breed, seed, feed, and weed," for short. The culture of trout, carp, catfish, and some variants of shrimp rearing are examples of such full-culture operations. Culture may also be partial; two varieties of partial culture are stocking seed without attending to subsequent growth at various levels of intensity, and caring for the animals to various levels of their life cycles. Cod grounds have been seeded—at least, attempts have been made in this regard. Salmon ranching that involves care of the fish in captivity to the fingerling or smolt stage belongs in this category, as does seeding shrimp larvae in Japanese waters. Conversely, one may collect the larvae or juveniles in the wild and rear them to harvesting size. Much of the mollusk culture, like that of scallops, is still of this incomplete-culture variety, as is the culture of certain fishes, such as Japanese yellowtail and some Southeast Asian milkfish, and most shrimp culture operations.

It is clear that increasing production through intensifying culture taxes the carrying capacity of the environment in which the farming takes place. Concentrating many animals in a small space creates high oxygen demands and increases the concentration of waste products (mainly nitrates and phosphates) and increases

TABLE 1.2. Developing-Country Aquaculture Systems: Environmental Impact and Benefits for Producers

System[a]	Environmental Impact	Benefits
	Extensive	
1. Seaweed culture	May occupy formerly pristine reefs; rough weather losses; market competition; conflicts/failure and social disruption.	Income; employment; foreign exchange
2. Coastal bivalve culture (mussels, oysters, clams, cockles)	Public health risks and consumer resistance (microbial diseases, red tides, industrial pollution); rough-weather losses; seed shortages; market competition especially for export produce; failures and disruption	Income; employment; foreign exchange; directly improved nutrition
3. Coastal fishpond (mullets, milkfish, shrimp, tilapias)	Destruction of ecosystems, especially mangroves; increasingly noncompetitive with more intensive systems; nonsustainable with high population growth; conflicts/failures and social disruptions	Income, employment, foreign exchange (shrimp); directly improved nutrition
4. Pen and cage culture in eutrophic waters and/or on the bottom	Exclusion of traditional fisheries; navigational hazards; conflicts and social disruption; management difficulties; wood consumption	Income; employment; improved nutrition
	Semi-intensive	
1. Fresh and brackish water ponds (shrimp and prawns; carps, catfish, milkfish, mullets, tilapias)	Fresh water: health risks to farm workers from water-borne diseases. Brackish water: salinization/acidification of soil/aquifers. Both: market competition, especially for export produce; feed and fertilizer availablity/prices; conflicts/failures and social disruption	Income; employment, foreign exchange (shrimp and prawns); directly improved nutrition
2. Integrated agriculture/aquaculture (rice/fish; livestock + poultry/fish; vegetables/fish; vegetables/fish; all combinations of these)	As for fresh water above, plus possible consumer resistance to excreta-fed produce; competition from other users of inputs such as livestock excreta and cereal brans; accumulation of toxic substances from livestock feeds (e.g., heavy metals) in pond sediments and fish; accumulation of pesticides in fish	Income; employment; directly improved nutrition; synergistic interactions between crop, livestock, vegetable and fish components; recycling of on-farm residues and other cheap resources

3. Sewage/fish culture (waste treatment ponds; latrine wastes and septage used as pond inputs; fish cages in wastewater channels)	Possible health risks to farmworkers and consumers; consumer resistance to produce	Income; employment; directly improved nutrition; waste disposal liabilities turned into productive assets
4. Cage and pen culture, especially in eutrophic waters or on rich benthos (carps, catfish, milkfish, tilapias)	As for extensive cage and pen systems above	Income; employment; directly improved nutrition
	Intensive	
1. Freshwater, brackish water and marine ponds (shrimp and prawns); fish, especially carnivores (catfish, snakeheads, groupers, seabass, etc.)	Effluent/drainage high in biological oxygen demand (BOD) and suspended solids; market competition, especially for export product; conflicts/failures and social disruptions	Income; employment; foreign exchange
2. Freshwater, brackish water and marine cage and pen culture (finfish, especially carnivores—grouper, seabass, etc.—but also some omnivores such as common carp)	Accumulation of anoxic sediments below cages due to fecal and waste feed buildup; market competition, especially for export produce; conflicts/failures and social disruption; consumption of wood and other materials	Income; foreign exchange (high-priced carnivores); a little employment
3. Other: raceways, silos, tanks, etc.	Effluents/drainage high in biological oxygen demand (BOD) and suspended solids; many location-specific problems	Income; foreign exchange; a little employment

[a]Extensive systems have no feed or fertilizer inputs, semi-intensive systems have some feed and/or fertilizer inputs, and intensive systems are mainly reliant on external feed inputs. Possible consequences of exotic breed transfers apply to all systems listed here. Enhanced fisheries are not included here because of the general lack of developing-country examples.

Source: Adapted from Pullin (1993). With permission from ICLARM and the author, R. S. V. Pullin.

likelihood of disease transmission. The use of various forms of chemical treatment against this latter contingency can pose public health problems.

Pollution by and of aquaculture can occur because of multiple demands on the surroundings of aquaculture operations. Local carrying capacities must often be increased by aeration if not by resorting to loading excess materials on the environment, trusting that they will be diluted. Dilution is being accepted publicly less and less, and some of these "externalities" tend increasingly to be internalized through various modes of recycling. These are prominent challenges for engineers and biologists.

Aquaculture may also be classified as extensive, semi-intensive, intensive, or even highly or superintensive, depending on inputs of energy, materials, and labor (Table 1.2). The classification shown in here does not connote size or extent of water area involved, which may be managed extensively or, in the case of several or many ponds, intensively. But it can imply social value and purpose. For example, family farms can be of the subsistence type, but they may also be managed intensively; often they are integrated with animal rearing.

The animal wastes and other farm wastes can serve as fertilizers for ponds but only when they are easily available with little or no transport, as with ducks on ponds in China. Thus the integration of rearing livestock and poultry with managing fish ponds offers ecologically attractive possibilities; it is done in China, Hungary, and now also in Africa (Pillay, 1996). But there may be certain public health risks connected with this type of culture (see Chapter 4). Larger aquatic farms or culture enterprises are most often semi-intensive or intensive and are undertaken only for profit. They can be of the partial-culture type when they are essentially feedlot operations for grow-out to commercial size of the cultivars. In the case of intensive mariculture, such as salmon in cages, spawning and early growth takes place in fresh water, separately from grow-out, while intensive freshwater cultures generally do not show this separation. Spawning, if done on the farm and not relying on centralized hatcheries, is usually done in the same water system as the grow-out. In contrast, aquaculture for purposes of conservation, usually including restocking of fry and fingerlings, is generally intensive and may rely on advanced technologies.

The era of introducing species into new environments, such as trout into New Zealand or the highlands of Africa in the 19th century, is not over, as the worldwide spread of Tilapia testifies, notwithstanding the present-day concerns about biodiversity (see Chapters 4 and 5). Many attempts are being made, however, to bolster stocks that have diminished, usually through overfishing. The challenge here is to plant endangered species fish that are of the size most likely to survive and to balance the cost of rearing them by the tens of thousands with socioeconomic returns; tagging released fish in a cheap and effective manner permits assessing that size (Leber, 1995).

Aside from full or partial life-cycle cultures, we may also categorize culture operations according to the biological attributes of the cultivars, the most important attribute being feeding types.

Feeding types among fishes range from predatory gulpers to sifters of organic materials in mud, to zooplankton feeders, to herbivores that eat algae or even leafy

TABLE 1.3. Proximate Composition of Various Feedstuffs Used in African Fish Ponds

	Percentages			
Categories	Carbohydrates	Fats	Proteins	Fiber
Animal remains: e.g., blood meal, salt fish waste	35–38	1–2	35–76	0–30
Corn products: bran, flour, dry stalks, grain	45–81	2–8	6–12	1–30
Other cereals: wheat bran, millet, rice, rice bran	60–80	3–4	5–9	1–23
Cotton products: cake, seed	30–39	7–19	33–48	10–25
Coffee, cocoa hulls	34–59	8–16	9–12	24–39
Legume products: ground soybean, peanut press cakes	27–31	8–16	34–54	5–6
Vegetable remains: lettuce, kale, pumpkin	4–6	0–1	1–4	0–2

Source: Selected from Miller (1975) and rounded to nearest integer.

plants. As already indicated, the rationale for polyculture is the selection of compatible species with different feeding patterns. In addition, because fish learn to feed on almost anything, it is relatively easy to develop pelleted food for fish culture, dietary quality considerations aside. At the same time, such catholic feeding habits permit the usẹ of plant materials, especially cheap or nearly valueless crop residues such as bran and millet (Table 1.3). The practice of building very wide pond margins to the fish ponds in China for cultivating grasses illustrates this; here leafy plant-feeding grass carp (*Ctenophryngodon idella)* comprise about 20% of the stock in the pond (Tapiador et al., 1976).

All sorts of other wastes, even sludge, are fed to fish with very low conversion efficiencies, to be sure; presumably this is done because it favors cheap production costs just the same (Table 1.4).

Matters are not as simple, though, as this discussion of categorizing culture operation according to feeding type would imply. This is because it is probably more difficult to make such a division of carnivores and herbivores in the aquatic realm than in the terrestrial realm, and aquatic omnivores also abound.

But zooplankton is also animal matter, and many plankton feeders like young salmon and browsers like shrimp require animal protein in their diets, as with rearing pigs and chickens, where keeping animals at high density simply requires extraneous feeding. In Chinese and Indian polyculture practices, where plant feeders of the carp family are important components, leaves and other green matter are supplied, and in other fishponds there is additional feeding of whatever surplus matter may be at hand (Tables 1.3 and 1.4). Intensive aquaculture of outright carnivores such as trout, salmon, seabass, Japanese yellowtail, and now increasingly also shrimp, to mention the most important ones, requires concentrated protein in the form of fish meal, unless discarded trashfish is at hand, as may be the case when groupers are reared near a shrimp fishery, like in Malaysia. This type of fish culture represents only 12% by weight, but far more in value, of all finfishes cultured in

TABLE 1.4. Yields of Fish for Various Residues Used in China

Residue or Feed	Residue or Feed Quantity (kg)	Fish Yield	Estimated Conversion Efficiency, % (kg yield/kg feed × 100)
Grass or vegetable tops	60–70	1 kg grass carp	1.4–1.7
Snails and clams	50	1 kg black carp	2.0
"Fertile water": 77% bean curd residue; 23% residue of fermented products	100	1 kg silver carp	1.0
Animal manure	25	0.5 kg silver or bighead carp	2.0

Source: Based on information given to mission members; after Tapiador et al. (1976); conversion efficiencies are our estimates.

1992 (Tacon, 1994). It uses approximately 15% of the world fish-meal production, in pellets that incorporate fishmeal to various degrees, least for catfish and most for the marine carnivores such as sea bream. Studies are under way to replace fish meal still further with alternative protein sources such as animal byproducts, single-cell proteins, oilseeds, soybean products, and the like (Tacon, 1994; see also Chapter 5).

Nearly 30%, or 27 MMT, of the total world fish catch of 1993 was composed of small, often oily, schooling fishes like anchovies (FAO, 1995), which are the basis of fish meals and industrial fish oils. There has been speculation about more direct use of these species for human consumption; they are in fact eaten where they occur when fresh, or they are canned (e.g., sardines and anchovies). But marketing them in volume directly for the table in other animal protein–needy parts of the world is likely to be technologically and economically impossible. But we must also consider that big fish eat little fish, so as long as fisheries bring the big ones to market, there will be culture for them; making fish meal just concentrates the small ones for easier distribution. Feeding pigs and chicken with fish (i.e., as fish meal) may be less natural than feeding fish with fish but it also will persist. In fact, with land animal husbandry far outstripping that for aquatic animals, in volume and importance, the contest for fish meal will also persist, and judgment is still out which will win.

For both land and aquatic animal rearing the trend in feed development of replacing fish meal with equivalent ingredients is important for their futures. This trend plays a role in the two overall development thrusts we can discern for aquaculture in the future: One comprises technical and managerial improvement of polyculture, integrated agriculture and aquaculture (now variously practiced in much of the world), and multiple use of water and recycling. The other, not necessarily strictly separate from the first, will be represented by new technologies, with genetics, biotechnology, and ocean engineering, leading to saving of water and efficient use of ocean space as some of its main ingredients. Extraneous feeding of fish cultivars is necessary here—all the better if this can be achieved in a socioeconomically and ecologically sustainable manner.

REFERENCES

Anonymous. 1995. Entwicklung and Zusammenarbeit. Deutsche Stiftung für Entwicklung. Berlin, 36(12):18.

Bardach, J. E., T. H. Ryther, and W. O. McLarney. 1972. Aquaculture: farming and husbandry of fresh water and marine organisms. John Wiley & Sons, New York.

Borgstrom, G. 1962. Fish as food, Academic Press, New York. 777 p.

Brune, D. E. 1994. Sustainable aquaculture systems. Report prepared for the Office of Technology Assessment, U.S. Congress, Food and Renewable Resources Program, Washington, D.C.

Chikuni, S. 1987. The fish resources of the Northwest Pacific. FAO Fisheries Tech. Paper 266, Food and Agriculture Organization of the United Nations, Rome.

Costanza, R. 1991. Ecological economics: the science and management of sustainability. Columbia University Press, New York.

Edwards, P. 1993. Environmental issues in integrated agriculture; aquaculture and wastewater-fed fish culture systems; environment and aquaculture in developing countries. pp. 139–170, In: R. S. V. Pullin, H. Rosenthal, and J. L. Maclean (Eds.). ICLARM, Conference Proceedings 31. Manila, Philippines.

FAO. 1981. Atlas of the living resources of the seas. Fisheries Series 15. Food and Agriculture Organization of the United Nations, Rome.

FAO. 1991. Environment and sustainability in fisheries. COFI/91/3. Document presented at the 19th Session of the Committee on Fisheries, April 1–12. Food and Agriculture Organization of the United Nations, Rome. 23 p.

FAO. 1992. Fish and fishery products: world apparent consumption statistics based on food balance sheets (1961–1990). Fisheries Circular 821, Rev. 2. Food and Agriculture Organization of the United Nations, Rome. pp. 25–53.

FAO. 1995. Aquaculture Production Statistics, 1984–1993. Fisheries Circular 815, Rev. 7. Food and Agriculture Organization of the United Nations, Rome. 186 p.

FAO. 1995a. The state of world fisheries and aquaculture. Food and Agriculture Organization of the United Nations, Fisheries Department, Rome. 57 p.

FAO. 1995b. Fisheries statistics, catches and landings. FAO Fisheries Series 76. Food and Agriculture Organization of the United Nations, Rome.

Gulland, J. A. 1974. pp. 68–126, In: The management of marine fisheries. University of Washington Press, Seattle.

Jhingran, V. G., and B. K. Sharma. 1980. Integrated livestock—fish farming in India. pp. 135–142, In: R. S. V. Pullin and Z. H. Shehadeh (Eds.). Integrated agriculture-aquaculture farming systems, ICLARM Conference Proceedings 4. Manila, Philippines.

Katz, A. 1995. Summary report: study of national strategies for aquaculture development. Meeting highlights: Sustainable Aquaculture. PACON '95, Hawaii.

Leber, K. M. 1995. Significance of fish size at release on enhancement of striped mullet fishery in Hawaii. J. World Aquacult. Society, 26(2): 143–153.

Miller, T. W. 1975. Fertilization and feeding practices in warmwater pond fish culture in Africa. FAO/CIFA Symposium, Aquaculture Africa, Accra, Ghana. CIFA/75/SRA. Food and Agriculture Organization of the United Nations, Rome.

Pauly, D., and V. Christensen. 1995. Primary production to sustain global fisheries. Nature (London), 374: 255–257.

Pillay, T. V. R. 1996. Economic and social dimensions of aquaculture management. Aquacult. Econ. Manage. (inaugural issue, PingSun Leung, Ed.). Honolulu.

Pullin, R. S. V. 1993. Discussion and recommendations on aquaculture and environment in developing countries. In: R. S. V. Pullin and Z. M. Shehadeh (Eds.). ICLARM Conference Proceedings 31. Manila, Philippines.

Ruddle, K. 1993. Impacts of aquaculture development on socioeconomic environments in developing countries: towards a paradigm for assessment. In: R. S. V. Pullin and Z. H. Shehadeh (Eds.). ICLARM Conference Proceedings 31. Manila, Philippines.

Tacon, A. G. J. 1994. Dependence of intensive aquaculture systems on fishmeal and other fishery resources—trends and prospects. FAO Aquacult. Newslett., 6: 10–16. Food and Agriculture Organization of the United Nations, Rome.

Tapiador, D. D., H. F. Henderson, M. N. Delmendo, and H. Tsutsui. 1976. Freshwater fisheries and aquaculture in China. FAO Aquaculture Mission to China, April/May 1976. FAO Fisheries Tech. Paper 168. Food and Agriculture Organization of the United Nations, Rome.

Vitousek, P. M., P. R. Ehrlich, A. H. Ehrlich, and P. A. Matson. 1986. Human appropriation of the products of photosynthesis. BioScience (Washington, D.C.), 36(6): 368–373.

2 Aquaculture and Sustainable Food Security in the Developing World*

MERYL J. WILLIAMS

INTRODUCTION

Despite, and sometimes because of, endeavors to transform and develop the biosphere for human ends, many of the world's poor and low-income people still lack reliable access to enough food to sustain their health and normal daily labors—they lack food security.[1] The absolute numbers of food-insecure people are growing annually, although at a global scale the percentage of people living in poverty is shrinking. In 1993, 1.3 billion people were classified as "the absolute poor," and 800 million people as not having sufficient and regular supplies of food (World Bank statistics, quoted in Commission on Global Governance, 1995).

Since this book is about one sector of economic activity, namely aquaculture, there is a necessary focus on the sector itself, including a bias toward the producers. However, I have tried to avoid a narrow sectoral approach and to highlight how the sector serves consumers, and how it relates to and is affected by other sectors and by trends and events in the larger general social, economic, natural, and political environment. Aquaculture development can be viewed as a special part of rural development.

*I am indebted to Arlene Garces of ICLARM for her contributions to the section on "Opportunities and Constraints" and to Dr. R. Brummett, also of ICLARM, for the use of material on African aquaculture from our joint paper in preparation. Dr. W. R. Hansen provided invaluable assistance in the preparation of the final manuscript.

[1]Food security is defined as "physical and economic access, by all people at all times, to the basic food they need" (AGROVOC, FAO's thesaurus used for AGRIS). Food security therefore embodies stable, sustainable, and predictable food supply, equity through access for all (though access to the means of production and/or purchasing power), and quality including nutritional adequacy for life functions. Speth (1993) noted that sustainable food security "fuses the goals of household food security and sustainable agriculture," therefore embodying the aspects listed and "the protection and regeneration of the resource base for food production—terrestrial, aquatic and climatic."

Sustainable Aquaculture, Edited by John E. Bardach
ISBN 0-471-14829-6

From the end of World War II to the late 1960s, development had an urban bias and was concerned primarily with economic growth and industrial modernization (Auty, 1995; Hettne, 1995). By neglecting rural development, governments overlooked the highly significant rural producers and consumers.

In the late 1960s and early 1970s, development models began to change, environmental awareness grew, and the green revolution in agriculture began to bear fruit. The green revolution demonstrated that farmers would adopt new technologies, provided the risks appeared acceptable, and therefore that rural investment could be viable (Auty, 1995). The incorporation of ecological sustainability into agriculture and rural development is now receiving attention (Breth, 1996). This is also the case in aquaculture, as demonstrated by the present book. In addition, women's roles and the increasing participation of civil society in development are getting greater attention in development.

SUSTAINABLE FOOD SECURITY

Sustainable food security depends on the sustainable supply of food, access to that supply, and its nutritional adequacy. Sustaining food supply requires protecting the environment as the basis for production. In the face of growing food demands from increasing urban and rural populations, the supply of all foods is important, and setbacks in any food production sector will place greater pressure on other sectors. However, most calculations of the adequacy of the food supply have focused on grains only since these are the staples of the diets of low-income people and any change in their availability signals changes in the overall food situation.

For global level consumption, increases in the average per capita supply and consumption of most plant foods (grains, cereals, legumes, roots, and tubers) has plateaued (Pinstrup-Andersen, 1994; FAO, 1994). Consumption of livestock products continues to increase, but consumption of fish and other aquatic products appears to have peaked in 1989 and to now be declining (Figure 2.1).

Little has been written on the consequences of interactions in food availability between grains and animal proteins, including fish. Delgado (1995) reviewed studies on responses of fish consumption to income and price of fish in Africa. For sub-Saharan African countries, he found that fish consumption rose as income rose, and fell as the price of fish rose. In Côte d'Ivoire in West Africa, cheap imports of frozen pelagic fish and fresh West African beef were good substitutes, displaying symmetric cross-price elasticities.

Fish has made larger gains than other foods over the last four decades. The 1990 per capita level was approximately 1.7 times the 1961 level for the developing world and 1.5 times the 1961 level for the developed world, despite significant increases in population during the period. The increases came largely from natural fisheries resources and only more recently from aquaculture. Now, however, after more than four decades of increase, the contribution of fish to sustainable food security is undergoing a transition to increasing scarcity. The solution will include increased aquaculture production (Williams, 1996).

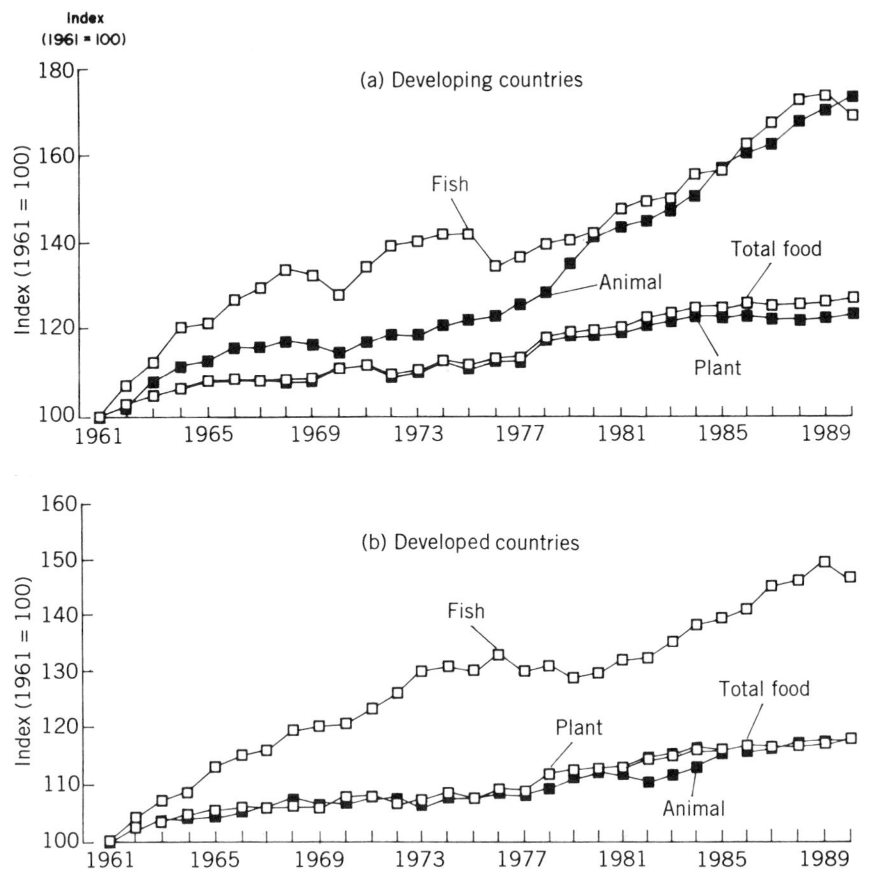

Figure 2.1. Daily per capita calorie consumption, 1961 to 1990. Reprinted with permission from the International Food Policy Research Institute (Discussion Paper 13), using data from FAO 1993.

Part of the transition has been characterized as "from hunting to farming fish" (ICLARM 1995). This part of the transition is under way, mobilized by rising demands for fish and improving technical knowledge. Since the Food and Agriculture Organization of the United Nations (FAO) first started collecting detailed aquacultural statistics in 1984 and until 1992, annual per capita fish supply from aquaculture increased by 67% (FAO, 1995a). Globally in 1992, aquaculture contributed an average of 2.5 kg of food per capita.

In December 1995, 95 nations met in Kyoto, Japan, to consider the issues of the contribution of fish to sustainable food security. This meeting represented a turning point in fisheries affairs as, for the first time, nations officially recognized "a potential shortfall by 2010 of the supply of fish and fishery products to meet demands from an increased human population, which in turn will adversely affect world food security" (Kyoto, 1995).

Predicting fish demand is complex. At Kyoto, the present per capita supply figures were used in projections, although as noted above these are a historical high. In the case of nutritious high-protein foods such as fish, demand is increased by economic factors and population growth. During the period 1970 to 1990, population growth accounted for only half the growth in fish consumption (Westlund, 1995). Economic and cultural factors such as greater disposable income, the price of fish relative to other animal proteins, trade opportunities, and dietary and health preferences all contributed to fish consumption patterns (Westlund, 1995). Many of these factors have acted positively on the demand for fish as well as on its price and tradability, thus putting pressure on all forms of supply and tending to limit access to it by poorer people. This is a break with the past in the developing world, when fish from natural stocks used to be considered "the poor man's protein" due to its low price and/or to the fact that the very poor who owned no land or other means of production nevertheless could often exploit open-access fisheries resources as commons.

With specific reference to aquaculture, documents presented at the 1995 Kyoto Conference projected that aquaculture production in 2010 would be between 27 million and 39 million tonnes (or metric tons, MMT), up from a 1993 total of 16 MMT (FAO, 1995b), thus providing the greatest potential hope for maintaining per capita fish supply. Marine and inland fisheries were predicted to stay steady, decline, or at best increase more modestly than aquaculture (5 MMT each as the most optimistic increases).

At an aggregate level, therefore, aquaculture is the major, though not sole, hope for improving the world's fish production. Few would deny that aquaculture will continue to make great progress as technology, know-how, and investment race ahead in the sector. But how will success in increasing aquaculture production help those who are food-insecure?

FISH AND FOOD SECURITY

Fish and all other aquatic products contribute to food security directly as nutritious human food supplying protein, essential amino acids not found in staples, and in some of their forms calcium (e.g., bones and scales in dried and canned fish), iodine, some vitamins, minerals, and other trace elements (Rogers, 1990). For developing countries, fish from aquaculture and fisheries combined currently supply 19% of total animal protein consumed and just over 5% of all protein (FAO, 1995b). Indirectly, but importantly also, fish assist food security by providing income for fishers and fish farmers, and livelihoods for workers in fisheries, aquaculture, fish processing, marketing, and allied supply and service industries such as feeds, extension, hatcheries, and so forth.

The three dimensions of food security—sustainable supply, access to food, and nutritional adequacy—are now each addressed.

Sustainable Supply Options

Fish supplies can no longer be left only to nature. Whereas most fish (85% of total fish or 79% of food fish) still comes from natural stocks, the majority of these are impacted negatively by heavy fishing (including overfishing), habitat degradation (including pollution), and habitat loss. Many management regimes are not adequate to sustain the resource base, and aquatic environments continue to suffer the downstream effects of most terrestrial and atmospheric events, including climate and climatic change. Even without the prevalence of negative impacts on natural fish production, the biophysical limits to natural production from natural stocks appear to have been reached or exceeded (Pauly and Christensen, 1995). Encouraging more people to catch fish for food or profit is no longer an option in most parts of the world. Moreover, encouraging people to grow their own fish or participate in artificial stock enhancement schemes are increasingly viable options.

Despite great technical advances, modern aquaculture is still a new technology and requires further progress to meet the supply challenges ahead. Enhanced development of aquaculture is only recent. Even though the aquaculture of carps began at least 2500 years ago in China, carps were only successfully bred in hatcheries as recently as the early 1960s. Most current aquaculture is still quite rudimentary, relying on natural supplies of seed stock, unimproved wild types of fish, and simple culture technologies and inputs. Feeds are also largely unimproved, and the nutritional requirements of most species are not known at all except in general terms from studies of diet and feeding preference.

The example of Norway's extraordinary success with the domestication and mass production of Atlantic salmon shows that the bounds to increasing supply through greater production efficiency are only now being explored. To some extent, the impetus to reach these bounds will come from the technology push itself. A greater impetus will be market demand for fish and local pressures for new forms of livelihood and enterprise.

The majority of aquaculture production occurs in developing countries, and production in the developing world has been increasing much more rapidly than that in the developed countries, due mainly to developments in Asia.

Have these rapid advances in supply benefited low-income people, and if so, how? Kent (1995) concluded that intensive aquaculture did not benefit the poor, and some forms may even have negative impacts. Extensive traditional forms are more likely to benefit the poor in developing countries.

Losses and shortfalls of supply from natural stocks are not directly replaced by aquaculture. The species showing the largest increases in aquaculture production are carps, tilapias, shrimps, and salmons. Shrimps and salmons satisfy luxury export markets and increasingly lower-priced markets in the developed world as cost of production drops (Anonymous, 1996), but not basic food needs. In Asian countries and in some North American export markets, tilapias are substituting for generic whitefish formerly supplied only by marine demersal species. Carps satisfy a range of low to middle price markets, especially in Asia, but tend not to replace either marine fish or small native freshwater fish in markets. There is virtually no export

TABLE 2.1. Number of Species Used in Aquaculture

	Number of Species	Percentage of All Species
Extant finfish	24,618	100.0
Fish used in industrial and artisanal fisheries	2,576	10.5
Fish used in aquaculture[a]	179	0.7
Fish used as bait	134	0.5
Fish used in ornamental trade	1,980	8.0
Marine fish	546	
Freshwater fish	1,434	
Mainly artificially reared	773	
Fish used as sport fish	798	3.2
Total used by humans	4,572	18.6
Finfish affected by humans		
Threatened	770	3.1
Introduced[b]	221	0.9
Finfish affecting humans		
Dangerous[c]	437	1.8

Note: Although FishBase does not yet contain all species, the above statistics should already provide a reasonable estimate, since ICLARM has made an effort to include all species that are used by humans. The number for fisheries is underestimated because many species that are important in artisanal fisheries are not reported in the literature. The same is true for bait fishes.

[a]Species used for food or stock enhancement in commercial aquaculture.

[b]Species transferred to and established in another country.

[c]Species that are, for example poisonous or traumatogenic (causing adversely altered mental states).

Source: ICLARM (1994); Williams (1996).

market for carp, so most is consumed locally. Larger carps command higher prices than smaller species such as silver barb (*Puntius goniotus*) which is cultured in its native Thailand and Cambodia and in Malaysia, Brunei, Bangladesh, and Indonesia, into which it has been introduced.

The greatest supply gap for coastal people will be in marine fish for which, in Asia, Csavas (1994) estimates that only 1.2% is presently supplied by aquaculture but for which demand is high. Many desirable marine species are carnivores, and their cultured forms are likely to remain out of the reach of low-income people. Tilapias are a replacement in some countries, such as the Philippines.

Aquaculture production will directly assist the supply of fish for food-insecure people if they grow it themselves or indirectly if it is grown cheaply enough for them to purchase or barter. Given the wide range of species and types of aquaculture (Tables 2.1 and 2.2), different types of enterprises and different scales of enterprise will have different, often location-specific impacts on food security (Table 2.2). For example, commercial enterprises producing luxury fish or shellfish may have positive impacts on food security through providing some local employment and general

TABLE 2.2. A Classification of Aquaculture and Stock Enhancement Enterprises, with an Emphasis on Developing Countries

Type of Enterprise	Main Distinguishing Features	Operators, Labor	Disposal of Production	Examples in Developing Countries
1. Leased pond or other small waterbody	Use arranged with owner or government; inputs minimal and depend on access to credit; extensive to semi-intensive	Landless people; usually rural poor	Lessee's family and/or local market	Bangladesh (Ahmed, 1992)
2. Subsistence farm pond (micro-enterprise)	Very small scale; minor part of agriculture/rural system, which is often a mixed-crop system, crop-livestock system, or farming-fishing system; inputs small and local in origin, many from farm byproducts or waste; extensive to semi-intensive; sometimes natural reseeding	Small landholders' own family; specialized labor may be used for part of the culture cycle (e.g., pond construction, harvest)	Farmer's family, neighbor's, and sometimes local market	Malawi (ICLARM and GTZ, 1991); mussel and oyster culture, Philippines (Newkirk, 1993); rice-fish farming in Asia; giant-clam farming; extensive coastal shrimp farming with natural reseeding
3. Small aquaculture enterprise	Small-scale; aquaculture major part of enterprise; inputs and services largely purchased; some inputs imported from nonlocal suppliers; semi-intensive to intensive	Owner and family; hired labor, usually seasonal	Almost all product sold on local or more distant (e.g., urban) markets	Thailand mussel and oyster farming (Newkirk, 1993); milkfish farming in Philippines; carp polyculture in China
4. Large aquaculture enterprise	Large-scale requiring large capital investments and infrastructure; most inputs imported from non-local suppliers; usually higher technology inputs and services required; intensive	Managed as a commercial business; all labor hired; most labor specialized	All marketed commercially, usually in urban markets; may be traded internationally	Penaeid shrimp culture in Asia, Central and South America; seabass culture in Thailand; some milkfish culture in Philippines

(*continued*)

TABLE 2.2. (continued)

Type of Enterprise	Main Distinguishing Features	Operators, Labor	Disposal of Production	Examples in Developing Countries
5. Cage culture	Cage culture is major part of owner's livelihood; inputs and services largely purchased; some inputs usually imported from nonlocal suppliers; usually intensive	Owner and family; hired labor, usually seasonal	Almost all product sold on local or more distant (e.g., urban) markets	Milkfish and tilapia cage culture in lakes and Philippines; carp and goldfish culture in cages, Indonesia dams
6. Public sector–supported stock enhancement	Large production hatcheries produce seed for distribution; extensive to semi-intensive	Government-sponsored management and labor for seed production; private sector harvesting similar to natural fisheries	Local, urban, and sometimes export markets depending on species and method of catching of product	Shrimps in Shandong province, China
7. Traditional culture and stock enhancement systems	In use since before modern aquaculture began	Local households, often organized along traditional community lines	Community consumption, local markets; highly dependent on local and traditional ownership rights	Damming of natural depressions and drain-in-ponds: Egyptian *howash* systems, *whedos* in Benin, Cameroon, Togo; fish aggregation devices: simple refuge traps and brushparks in Asia (Bangladesh, Philippines, Cambodia, China, Sri Lanka), Africa, Pacific, and Indian Oceans on fringing reefs; more complex brushparks such as the *acadja*, Benin (ICLARM and GTZ, 1991)

8. Small-scale, community-based stock enhancement (modern)	Similar to no. 7 above but recent development or new in given locality; reservoirs, ox-bow lakes, coastal bays, and estuaries with community tenure; some inputs e.g. fish seed, purchased; operations usually extensive or at most semi-intensive	Local community; seasonal hired labor or specialized services	Product consumed in community households and sold on local, urban, and rarely export markets	Oxbow lake systems, Bangladesh; Sri Lanka reservoirs
9. Wastewater and sewage fish systems	Large-scale; often associated with municipal wastewater systems	Commercial operators; infrastructure extensive	Product consumed on local commercial markets	Hanoi, Calcutta (Edwards, 1996)
10. Dams and reservoirs	Large-scale; artificial stocking, often with introduced species; cages culture may be added	Commercial-scale but may be small- as well as large-scale operators	Product consumed on local commercial markets	Most countries, and an increasing potential source of fish production, as more natural water resources are impounded

economic gain. At the same time, the enterprises may have negative impacts through physically displacing small-scale farmers or fishers who are those most likely to be living in poverty.

As can be seen from Table 2.2, aquaculture and stock enhancement already provide human food, employment, and income through a myriad of methods, systems, and species. In addition to these enterprises, there are ancillary services and enterprises that are needed to support aquaculture production. These include hatcheries, fingerling grow-out, fry collection from nature, feed processing, pharmaceuticals, specialist engineering services and suppliers, harvesters, postharvest processing, marketing, and storage.

Access

During the 1980s, studies on poverty and hunger shifted their emphasis from food-supply-based analyses to income-based analyses (Drèze and Sen, 1989). This shift is relevant to understanding how aquaculture may contribute to improving access to food security because the sector produces a highly marketable product. Nevertheless, aquaculture is also a direct supplier of food to those who culture fish.

Many studies have shown that aquaculture gives direct access to the household food supply in those low-income households that begin to grow their own fish and, depending on proximity to markets and quantities of fish produced, also contributes to improving their income and/or barter trade. Where producers are close to urban or local community markets, and as the market price appreciates, more fish will be sold for cash, making fish and aquatic products fully commercial rather than subsistence or artisanal staples.

Much of the fish grown by low-income farmers is eaten at home. Gupta et al. (1992) showed that fish produced by low-income families in Bangladesh increased animal protein consumption because 70% of fish produced was consumed in the household. Gupta and Rab (1994) found that about half of the fish grown in leased ponds by landless people in a study in Bangladesh was eaten in the home, the rest being sold at the local market. Aquaculture thereby improved access to food by direct growing of fish and by improving income to purchase foods and other needs such as housing, education, and medical services. In a study of 1200 farms in Guatemala that added a farm pond to land holdings of 2 ha (hectares) and less of land, about half the production (23 kg/yr per 120-m^2 pond) was eaten in the household, 20 kg was sold and 5 kg were used for restocking or given to neighbors (Popma et al., 1995). The household consumption of fish increased from 0.5 kg per person to 3.3 kg per person as a result of growing fish; and household income increased by 18% from the sales (Popma et al., 1995). In Malawi, small-scale fish farmers frequently barter fish for other food because markets are often distant from rural households, and affordable and reliable transport severely limits market access (Brummett, 1995).

Aquacultural types 1, 2, 6, to 8 in Table 2.2 seem therefore to definitely lead to greater direct access to food, including fish for home consumption by poor people. These categories may be where attention could be focused for aquacultural development assistance if food security outcomes are desired, although we should not assume that aquaculture is the only type of enterprise that could assist. Aquaculture may bring additional benefits if it enables households to rise above subsistence farming levels and into more commercial scales of operation, not necessarily only in aquaculture (Brummett, 1994).

Aquaculture also can improve access to food by those who labor for wages on larger enterprises and in ancillary activities and who thereby gain income.

More difficult questions concerning how aquaculture affects access to fish as food are (1) whether aquaculture will bring the cost of producing fish down sufficiently to allow low-income people who do not grow their own to purchase it, and (2) whether aquaculture is a better food-securing technology than others, for a particular place and time. These will be particularly important questions in those areas where natural fish supplies are or were the mainstay of local animal protein diets and where these supplies are no longer adequate. In these countries or locations, aquaculture tends to receive the most attention because the focus is on replacement or complementary supplies of fish. Candidate developing countries are Bangladesh; most Southeast Asian countries and parts of the Indo-Chinese countries surrounding the Mekong River, its tributaries, and the Great Lake in Cambodia; Malawi, parts of many other sub-Saharan African countries surrounding the Great Lakes, Senegal, Ghana, Nigeria, and Egypt; and Indian Ocean, Pacific, and Caribbean island countries.

For low-income people who cannot catch or grow their own fish in these countries, affordable new products are essential, but these will not always be direct replacements for traditional species. Cheaper fish are most likely to come from species with short production cycles, which require low inputs and therefore little capital investment. Omnivores and herbivores (carps, barbs, and tilapias) are good candidates. Production of these species has increased dramatically over the last decade in Asia. However, the increasing availability of seed for these species and the rising price of fish could cause their larger scale and more intensive cultivation; this could provide greater production but also greater incentive for more expensive, value-added products that are less affordable and/or less available on local markets. The extreme variation in prices of some products from the same species depending on market niche is a phenomenon almost unique to fisheries products. For example, there is a great difference between the prices of canned tuna and sashimi tuna.

However, aquacultural activities are not purely additional to all other forms of rural food production, employment, and income generation. The net impacts of aquaculture on food security have yet to be investigated in any specific location.

On land there is a finite amount of space available, which creates competition for it access. Aquaculture is often more profitable than farming alternate staples such as rice. Fish ponds therefore are starting to encroach on existing agricultural land at a time when these lands are at a premium. For example, in China, growing freshwater crabs (*Echiochiera chinensis*) produces 20 times the value of the same land area sown to rice (Li S., 1995). Fish farms are reported to be taking over agricultural land, encouraged by government policy "to create wealth from aquaculture" (Tyler, 1995).

The forms of aquaculture most likely to cause negative as well as positive impacts on food access are those producing products of high value or at least higher value than prior land uses and those that result in habitat destruction. These often displace existing uses, and rarely do high-value enterprises remain in the control of low-income people, even when the costs of production are low. Penaeid shrimps, milkfish (*Chanos chanos*), sand goby, and carnivorous fish for the live restaurant trade are examples.

Shrimp culture has received prominence because it has been one of the fastest growing forms of aquaculture in the developing world. In just over 10 years it has risen from negligible production to supplying about 800,000 metric tons, or a third of total world shrimp production. The coastal tropical and subtropical range of the major penaeids cultured has made this kind of aquaculture predominantly a developing-country activity targeted at export markets in the developed world. However, most of the sites most suited for shrimp farms were already sites for numerous other coastal activities, often but not solely associated with mangrove ecosystems and almost exclusively exploited by the poor for fishing, farming, and forestry.

In the coastal zone, removal of large areas of mangroves, particularly during the 1980s and early 1990s in Southeast Asia and southern Asia, to make shrimp and milkfish ponds has negatively affected coastal rice fields and fisheries and left large swaths of the coast unprotected from typhoons and storm surges (Phillips, 1995). The results have frequently led to conflict between the interests of the wealthier shrimp farmers and the poorer rural inhabitants. One observer commented that, in India, "shrimp farms have come to symbolize islands of prosperity in a sea of rural poverty" (Jayaraman, 1995). Most of the shrimp produced is exported from the local region, often to international markets. Frequently, the shrimp farm owners are also based in the cities; thus, income from the farms also leaves the rural economies. Many farm inputs are purchased from outside the area.

The situation in the Philippines illustrates the divide between the economic scales of enterprises. Under the government's Comprehensive Agrarian Reform Law, fishponds and prawn (shrimp) farms were specifically exempted by a major amendment in the sixth year of a 10-year program of redistribution of various tracts of land to farmer beneficiaries (Villanueva, 1994). The reasons given were that these aquacultural industries are highly technical and capital-intensive and small holders could not benefit from redistribution of such land. Subsequently, however, provision was made by the Philippine government to give priority to small farmer cooperatives and organizations when approving the conversion of public agricultural lands to fishponds and prawn farms.

Similar conflicts between shrimp farming operations and other rural activities have been reported from other localities such as the eastern Indian coast (Jayaraman, 1995), Malaysia (Hiebert, 1995), and Java, Indonesia (Bailly and Malistyani, 1996).

On the other hand, culture of some other high-value species such as giant clams, some algae, and some parts of pearl farming enterprises are feasible for low-income people, especially those who hold household or community tenure over marine territory. Inputs for these enterprises are inexpensive and therefore within the range of rural people. The returns on investment are high.

Human Nutrition

Per capita supply statistics are averages and therefore do not indicate the distribution of food among households. Inadequate nutrition is rarely caused by lack of supply of a single commodity such as fish but rather by poverty or inadequate income, some cases of which can be due to falling supplies and prices of fish.

Aquaculture should bring higher income and more animal protein into the household, at least for those households that retain their access to the means of production. However, studies on the commercialization of agriculture in many countries have shown that increased household income does improve nutrition but more slowly than expected (Kennedy and Bouis, 1993). Health, education, and sanitation programs seem necessary for families to benefit fully. Where capture fisheries occur, the nutritional adequacy of aquacultured fish may depend on whether it takes the nutritional place of the fish that were traditionally eaten. Culture often focuses on large, fast-growing species such as carps and larger tilapias. The traditional fish eaten in much of Africa and parts of Asia are small and eaten whole, often dried. The bones, scales, and skin of these fish provide valuable dietary calcium and other minerals (Deelstra et al., 1994), elements lacking in the muscle of a large fish.

THE DISTRIBUTION OF BENEFITS AND POLICY

Adoption of any new technology or variant of a previous technology will potentially benefit many people, but technological benefits are never equitably distributed and rarely totally free of risks; e.g., see Conway and Barbier (1990) and David and Otsuka (1994) on the outcome of the so-called green revolution in agriculture.

The issues relating food security and aquaculture indicate that the benefits of increased aquaculture adoption and production, sometimes called the "blue revolution," will not be equitably distributed. Some forms of aquaculture are only accessible to commercial operators and may even disadvantage poorer people. Capital and labor costs, technical skills, and the financial security to bear the risk of a new enterprise restrict many from venturing into some forms of aquaculture.

National policy objectives are important to the development of aquaculture and can have major impacts on the distribution of benefits. In all countries where aquaculture is judged to be succeeding, national governments have identified it as a priority for national interests, be they for export earnings, import substitution, food supply, technological advantage, or national security through decentralizing industries into remote areas (Katz, 1995).

Csavas (1994), drew attention to the tension between six pairs of common national policy objectives found by an FAO review of national aquaculture policies in the late 1970s and early 1980s (Table 2.3). All of these policy objectives influenc-

TABLE 2.3. National Policy Objectives for Aquaculture in Asia

Pairs of Contrasting Objectives	
Highest volume of edible products	Highest income for producers
Highest foreign exchange earnings	Import substitution
Highest absorption of labor	Highest productivity of labor
Highest utilization of resources	Conservation of resources
Mobilization of private investments	Assistance to small farmers
Development of vertical integration	Community development

Source: Csavas (1994).

ing aquacultural development can have positive contributions to food security, but some will not ensure it. For example, income generation alone will neither deliver the best outcome for food security in the developing world, nor ensure long-term environmental sustainability. Policy interventions must contribute collectively to sustainability and equity as well as profitability. In addition to government policies, market demand for fish, private-sector investment, and institutional factors affect how aquaculture will develop and how the benefits will be distributed.

Regional Analyses

Production statistics show that aquacultural development has been very uneven across regions and between the developing and developed world. From 1984, when detailed statistics were collected, until 1993, developed-country production grew by 30% (to 3.33 MMT from 2.55 MMT), whereas developing-country production has grown by 177% (from 4.8 MMT to 13.3 MMT) (FAO, 1995a).

In 1993 by continent, Africa produced 0.4% of aquaculture production, Asia 85.8%, South America 1.5%, North America 3.5%, Europe 7.3%, the former USSR 1.1%, and Oceania 0.4%. The global average of aquaculture production as a percentage of total food fish supply was 14% in 1992; this breaks down to 24% for Asia, 1.9% in Latin America and the former USSR, and 1.3% in Africa. Many major developing-country producers have doubled (e.g., China, India, and Thailand) or nearly doubled (e.g., Brazil, Ecuador, Philippines, and Vietnam) production since 1984. In Asia, only North Korea has declined as a major producer. Production has increased in all major groups, especially carps, milkfish, tilapias, penaeid shrimps, scallops, and mussels. Developing-country production of freshwater carps and barbs increased by 227% from 2.4 MMT in 1984 to 5.5 MMT in 1991.

Information on aquaculture production is uneven across regions. In Africa, for example, the bulk of North African production was estimated by the FAO rather than directly reported by countries until 1990; in sub-Saharan Africa, the percentage estimated has increased since 1988, thus making actual production figures uncertain (FAO, 1995a, 1995b).

In all regions, however, small-scale and subsistence production are the most difficult to estimate, and much is likely to be omitted altogether from published statistics. Since most freshwater fish in most regions are grown in earthen ponds, occasional estimates of the number of ponds give some estimate of aquaculture activity but are very inadequate for estimating production. The regional analyses that follow are therefore based on research reports rather than production statistics.

Asia. The first and major beneficiaries in Asian aquaculture have been the middle to large-scale commercial farmers and the consumers. Nevertheless, powered by growing industrial economies, the successes of technology, strong market demand, the ready availability of many inputs such as fish seed and fingerlings, and the popularity of easy-to-grow species such as carps and tilapias, Asia has also produced many small-scale winners and could produce even more. For example, studies in progress in five Asian countries (Bangladesh, China, Philippines, Thailand, and Vietnam) show that elasticity of fish demand with price varies across socioeconomic groups. In Bangladesh, where fish has traditionally been an important animal protein, lower-income people would eat more fish if the price were lower, whereas demand by higher-income groups is inelastic to lower prices (M. Dey, ICLARM, personal communication, 1996).

In many Asian countries, the gap between urban wealth and rural poverty is widening as industrial development is fueled by the industrial and information technology revolutions. A few forms of rural endeavor, however, also benefit from urban growth. Commercial aquaculture is one. It is driven by growing affluence as well as growing population numbers. It usually requires high capital inputs, technical know-how, and ownership of or access to land and water or coastal space. It is usually not an option for low-income people and, as shown by the shrimp farming example, it can be more lucrative than other forms of agricultural land use. Nevertheless, the efficient production of lower-value species such as carps and tilapias will put them within the reach of more low-income consumers.

The most important producer options for low-income people are in rural systems these people control themselves, i.e., those Newkirk (1993) described as "peasant microenterprises." Four are discussed here for Asia: integrated aquaculture-agriculture (including rice-fish culture), coastal aquaculture, use of natural and manmade water impoundments, and ancillary activities. Although comparative statistics are not available, it is likely that Asian small-scale producers are far more numerous than those in other regions, given the numbers of rural, (including coastal) people and the traditional popularity of fish.

Integrated Aquaculture-Agriculture (Including Rice-Fish Culture). On land, agriculture is the dominant land use in Asia. Aquaculture can be integrated into many different farming systems via use of multipurpose farmponds and other water sources. Economic and environmental benefits ensue, including the production of fish, improved recycling, better on-farm natural resource management, and the ability to spread financial risk through farm diversification.

The best-known form of integrated aquaculture-agriculture and one with a long history in parts of Asia is raising fish in rice fields and in ponds on rice farms. This is part of the popular image of Asian agriculture. Until the 19th century, this activity was more a feature of Chinese farms and of those in Bengal (West Bengal of the present India and Bangladesh) and Kerala regions of southern Asia, rather than generally widespread throughout Asia. Fish farming was rare in Southeast Asia, inland fish farming having been introduced into most countries by Chinese immigrants in the late 19th and early 20th centuries (Le Thanh Luu et al., 1995). Coche (1967), quoted in Ali (1992), sourced earlier introductions in Southeast Asia to India about 1500 years ago. One of the few Southeast Asian indigenous systems of rice-fish culture is based on a strain of the common carp (*Cyprinus carpio*) in northern Vietnam and northern Laos (Le Thanh Luu et al., 1995). According to Edwards and Little (1995), probably fewer than 10% of Southeast Asian farms culture fish even today.

In China, rice-fish culture includes fish raised in rice fields, fish in rotation with rice, and fish in waterways supplying irrigation water to the rice fields. Even in China, where rice-fish culture is nearly 2000 years old (Cai et al., 1995), fish culture has never been as widespread as at present; modern approaches to improve rice-fish production have been applied only during this century. During the early 1960s to mid 1970s, rice-fish production declined drastically as rice production intensified and chemical insecticides were heavily applied. From the mid-1970s on, however, less toxic chemicals, better varieties of rice, and the upsurge in availability of fish seed as aquaculture developed, all led to renewed interest in fish production from rice fields. Between 1981 and 1986, the area of rice-fish culture in China rose from 121,000 ha to 988,000 ha; and the number of provinces and municipalities reporting rice-fish culture rose from 4 in 1981 to 18 in 1986 (Cai et al., 1995).

Across nine Asian countries,[2] out of a total of 114.3 million ha of rice-field area, 10.2 million, or nearly 10%, are thought to be suitable for rice-fish culture. Far less than this, however, is currently used. For example, only about 20% of China's suitable area was estimated to be used in 1986 (Lightfoot et al., 1992).

Through rice-fish culture, a large number of small rural producers are benefiting from aquaculture in their rice fields. The fish produced are either fingerlings for on-growing in other aquacultural systems or table fish for domestic market and household consumption. In one study in the mid-1980s, Lin et al. (1995) reported that rice-fish culture led to financial returns that were 41% higher than rice alone. Part of the increase was due to increasing rice yields as a result of the apparent mutualism of rice-fish production; i.e., adding fish to rice fields tends to increase, or at least not decrease, rice yields. A review across Asia in 1988 showed few cases of decreased yield and even then the declines were very small (Dela Cruz et al., 1992).

Rice-fish culture is an important form of integrated aquaculture enterprises, especially in Asia, although it does not demonstrate the full range of features possible in the technology, such as the recycling of nutrients from ponds to fruit, vegetable, terrestrial livestock, and fodder production and of animal and plant waste into pond fertilizers. The concept of integrated resources management has been

[2]Bangladesh, China, India, Indonesia, Korea, Malaysia, Philippines, Thailand and Vietnam; see Table 1 (p. 2) from Lightfoot et al. (1992).

developed to encompass not only integrated aquaculture-agriculture, but other forms of resource flows within farming systems (Lightfoot et al., 1993).

It seems that the benefits of integrated aquaculture-agriculture often come more from the nonfish components of the farm, particularly from vegetable production. Prein (1966) showed in Ghana that 95% of the added economic benefits of a farm pond came from vegetables and only 5% from fish. The environmental benefits of recycling and farm diversification could not be quantified, but there are also nonfish benefits of having a farm pond.

Integrated aquaculture-agriculture has a patchy record of development, despite benefits that have been demonstrated experimentally. The potential use of integrated systems in Asia is enormous, including on small family farms; extensive integrated shrimp farming in mangroves systems; small-scale garden-pond-livestock pen systems; and large-scale industrial enterprises involving pigs, chickens or ducks, and fish as well as field crops, fruit, and vegetables. Larger systems can be fully commercial enterprises, almost all within reach of the more lucrative urban markets.

Recent studies in Australia have shown that diversified farms perform better in the long term because they are able to cope better with market and climate fluctuations than nondiversified enterprises (Kokic et al., 1995). However, the constraints to diversification, in terms of the economic investments and the knowledge needed, are high.

Most of the emphasis on integrating aquaculture into agricultural systems has focused on the landowner or its lessee. Work in Bangladesh has shown that, where small water bodies already exist, even technically landless rural people can take up aquaculture by leasing these water bodies, often with the assistance of nongovernmental organizations (Gupta and Rab, 1994). However, private and individually owned water bodies were more likely to have better fish-farming systems than jointly owned ones (Ahmed, 1992), and these owner operators were better off than the rest of the rural population (Ahmed et al., 1993).

Coastal Aquaculture Enterprises Throughout Asia, coastal space is limited and in demand from a great range of urban, industrial, tourism and other uses including fisheries. In monsoon, typhoon and cyclone-prone countries, natural seasonal hazards are particularly acute on the coastal fringe.

Culture of penaeid shrimps, except in some extensive forms integrated with mangroves, has been more of a detriment than a benefit to food security. Likewise, brackishwater pond culture of milkfish has had mixed outcomes for food security. As with shrimp farming, large areas of natural mangrove have been cleared to construct the ponds. In the Philippines, Chong et al. (1984) described the dualistic nature of milkfish farming wherein a minority of farms were extensive and used no inputs of fertilizers. Among the intensive farms using fertilizers, inputs, and therefore the productivity of the farm, were often restrained by their costs, the market value of the fish and the farmers' willingness to learn from other farmers. The profits from milkfish farming are therefore much better for those who start with greater economic means.

Mollusk culture, chiefly of bivalves, has been the dominant form of small-scale coastal aquaculture in Asia. More recently, seaweed culture is also becoming more important.

Despite the large number of edible mollusks in Asia, only a small number are cultured and most are limited to areas with natural spatfalls (McManus, 1995). Throughout most of coastal Asia, blooms of toxic dinoflagellates increased during the 1980s, placing major seasonal constraints on shellfish culture. In acute situations, seasonally affected coastal communities in the Philippines have been given "food aid" (Anonymous, 1995).

Seaweed culture has taken over from the collection of natural resources in some Asian countries. In the Philippines, the aquaculture of *Eucheuma*, *Gracilaria*, and *Caulerpa* now derives totally from cultured stands, and production is increasing, chiefly for export. It is seen as a good industry for coastal communities since it requires low inputs, and provides good returns and employment because it is labor intensive (Pagdilao et al., 1993). In parts of Indonesia, however, culture is limited by its remoteness from markets and by the people's lack of technical knowledge. Therefore collection of natural resources still predominates (Hatta and Purnomo, 1994).

Further development of various forms of coastal aquaculture suitable for small-scale operators is possible but will require considerable special investments in research and development, including into technologies, and in understanding local cultural and economic circumstances. McManus (1995) has pointed out the limited coastal environment of suitable quality (depth, currents, and water quality). He suggests that production of high-value products, rather than high-volume, low-value ones, should be targeted, but only as part of integrated coastal and national development programs, including proper management of coastal fisheries.

Lakes, Dams, Reservoirs, and Other Water Bodies. As more water is being impounded for urban, agricultural, and industrial use, natural aquatic systems are being disrupted, but new aquacultural and stock-enhanced fisheries potentials are also being created. In China, for example, the water area in reservoirs accounts for 40% of the total inland water surface area (Li and Xu, 1995). Natural lakes also provide fish-culture opportunities, albeit with risks to local biota.

Cage culture, floating net culture, and fenced or bay culture are all possible with suitable institutional and technical support. De Silva (1995) has proposed that reservoirs and lake systems have a much greater potential than many other aquatic resource systems for productivity increases. He bases his conclusions on the increased volumes of water being impounded, their presently underestimated production, the relative ease with which they can be stocked, and the ready availability of technology for their stocking.

A World Bank project involving the Saguling and Cirata reservoirs in western Java (Indonesia) in the late 1980s demonstrated that job opportunities for the rural people displaced could be created in fisheries and aquaculture with the appropriate training and early financial assistance. The report of this project warned, however, that these and nonagricultural job opportunities would not happen without assistance, since the rural people displaced initially lacked the financial means and training (Soemarwoto, 1990).

Ancillary Employment. Service industries for large-scale aquaculture are typically full commercial enterprises and have little direct benefit to low-income people. In Asia, notable exceptions are in service or allied industries where small operators have created niches. The enterprises that have developed are location-specific. Three examples are described below: fry collection from the wild, hatchery-to-pond operations in Bangladesh inland aquaculture, and postharvest processing.

In both shrimp and milkfish farming, small-scale industries have developed to collect natural fry or postlarvae along the shore. While creating employment for many poor people, these activities are at best very seasonal, and usually temporary until hatcheries become established. They may also be threatened by coastal environmental pollution, degradation of the habitat, and overexploited breeding stock (Castro, 1995). Collection of *Penaeus monodon* (giant tiger prawn) postlarvae in Bangladesh employs many thousands of poor fry collectors, but it also results in the destruction of 99 zooplankton individuals for each *P. monodon* postlarva collected, many of these are even fish larvae (Mahata et al., 1995). Since hatcheries are unlikely to be established in the near future, the collectors are being taught how to better sort the collections and release the non-target specimens live.

In Bangladesh during the 1980s, small-scale fish hatcheries producing carps, tilapia, and catfish seed developed to serve the needs of fishponds. There then developed a chain of small-scale employment involving up to 17 distinct activities to transfer the fry from the hatcheries to fingerling ponds, thence to sale and transport across country before purchase by pond operators. For example, young boys provided manual aeration, and women changed the water in fingerling containers en route.

Postharvest processing and marketing of fish is dominated by women in many Asian countries. Marketing chains for capture fisheries and aquaculture are often different due to the different sources of supply. In Bangladesh, Ahmed et al. (1993) found that a chain of traders and middle agents greatly reduced the profit margins of the capture fishers, but these have not yet developed to the same extent in aquaculture.

Pacific, Indian Ocean, and Caribbean Island Countries. Pacific, Indian Ocean, and Caribbean island countries, including many microstates, depend on seafood; their populations are growing fast, and their natural marine resources are often heavily exploited. On the face of it, this would seem a powerful combination of factors driving aquaculture. However, the resource situation and population pressures are relatively new, and extending fishing grounds and technologies, including fishing deep-water slope species, has managed to help fish supply meet rising demands until recently. Sporadic aquacultural efforts over the last three decades have been promoted by technological interests and often have not been tailored to the circumstances and development needs of island people.

The needs and the approaches to development are changing. Aquaculture is now in demand from more rural communities, and its development is now viewed more as part of rural development than as a separate sector or even as a part of the fisheries sector. With still too few technologies on offer, however, aquaculture is still not an automatic and integral rural development option.

Thorne-Miller and Jaildagian (1995) have proposed three principles for sustainable aquaculture development in the Caribbean: diversification (farming a variety of species both in polyculture and over any region), integration with the ecosystem and other socioeconomic activities, and localization (meeting local needs, including nutritional needs).

By paying careful attention to local human and biological resources, markets, and feasible technologies, several successes have been won in the culture of high-value invertebrates for export markets, notably for pearl oysters in French Polynesia and the Cook Islands (Gervis and Sims, 1992) and for giant clams (Tridacnidae) in Palau and the Solomon Islands. New technologies being developed for seaweeds and for invertebrates such as holothurians (sea cucumbers), green snails, and trochus offer further hope.

To achieve food security, high-value species, preferably those that are low on the food chain, are readily processed, and are easily transported to distant markets, will be best (Munro, 1995). Aquaculture should have little or no negative ecological impacts and use affordable technologies so that low-income people can become producers. Few reef fish are economically viable in this regard unless close to markets such as tourist resorts. However, given the apparent large natural wastage of reef fish larvae, sea ranching through grow-out of live-caught juvenile reef fish may be feasible.

As for coastal environments, suitable culture environments must still be defined for many island and reef species. Marine tenure systems also govern culture development. Hviding (1993) has examined the case of Solomon Islands for giant clam culture. Access to most reefs in the country is regulated by customary marine tenure. He concluded that in one sense this limits access to aquaculture sites, but in another it provides secure rights to farming territory.

Coastal households in island countries are already involved in a multiplicity of microenterprises. Therefore, Hviding (1993) has concluded, most new aquaculture will be part-time. Bell et al. (in press) has found that those enrolled in village trials growing clams in the Solomon Islands are mainly farmers who devote about 25% of their time to clam farming operations.

The risks in new enterprises are high, due not only to lack of knowledge by farmers but also to the very novelty of the enterprises themselves. The domestication from scratch of new species requires extensive research and experimentation. Again we consider the giant clam example, noting that these species gain their nutrition by photosynthesis and thus do not even need feeds; early village trials showed survival rates of 20% to 80% before the factors governing successful grow-out in lagoons were better understood (Bell et al., in press). Until basic husbandry is understood, therefore, most small-scale rural farmers cannot afford the levels of risk required. Research investments are essential for development.

An account of the full benefits of island aquaculture will often include cultural and aesthetic values. Giant clams and other shells are powerful traditional symbols in many island cultures (Hviding, 1993).

Africa. The African continent is the most threatened by rising food insecurity due to steeply increasing population, large-scale climate uncertainty, and declining levels of external assistance. While capture fisheries production has stagnated throughout the continent at about 8 kg per person, aquaculture has continually increased from 50 g per person in 1984 to 100 g per person in 1992 (FAO, 1995a). However, this is still only 1.3% of total fish intake, and this on a continent where many have predicted great potential for aquaculture development, given the availability of suitable fish species and water resources. A recent study by Kapetsky (1994) found that 31% of the area was suitable for warm-water subsistence aquaculture and 9% of the land in sub-Saharan Africa is suitable for commercial aquaculture.

In most African countries, aquaculture would be called an emergent subsector (less than 1000 metric tons of production and few species produced) under Katz's (1995) scheme. A few, such as Egypt, Nigeria, and South Africa, could be classified in the established but simple subsector. Africa is all but invisible in worldwide commercial considerations. For example, a recent market survey of companies listed only one company in one African country (Zimbabwe) in its tables (Ratafia, 1995).

Commercial and Rural Development Roles. Aquaculture plays two important roles in Africa (Williams and Brummett, in press): commercial and rural development. Both forms have parallels in other regions, but the commercial aquacultural sector, like most other commercial sectors in Africa, is less developed than in any other region. Commercial developments focus on high-value species, usually those close to major markets or for export. In Africa, these enterprises are often short-lived due to the range of circumstances common to many agricultural industries including poor infrastructure, price fluctuations, and difficulty of obtaining essential inputs.

Much of the description that follows on the role of small-scale aquaculture in rural development also applies to all other regions. In Africa, however, the issues have received more attention because there this sector has not been overshadowed by rapid commercial aquacultural development. In Asia, commercial successes have taken the focus off small-scale development, although the small-scale sector's problems are very similar to those in Africa.

The two types of aquacultural development make different contributions to African economic development, and each has different needs. Development has shifted its emphasis over time between the two, as shown in Table 2.4.

Currently, about 97% of total aquaculture production comes from the rural development sector (King, 1993) and is produced by cash-poor small-scale farmers in small ponds supplied only with on-farm inputs such as brans, manures, and weeds. The overall performance of aquaculture is poor, and repeated reviews over the last decade (e.g., King and Ibrahim, 1988; Huisman 1986a, 1990; Harrison et al., 1994) have confirmed its slow progress.

TABLE 2.4. Milestones in African Aquaculture

1920s	*Development:* Fish culture starts in Kenya (1924).
1930s	*Development:* Fish culture starts in Zaire (1937).
1940s	*Development:* Fish culture intensifies in Zaire (1946+); starts in Zambia (1942), Cameroon (1948), Congo-Brazzaville (1949), and Zimbabwe (1950).
1950s	*Development:* Rapid development including an increase in the number of ponds, spreads to many countries. *Research:* Tilapia culture
1960s	*Development:* Spread and development of fish culture peaks; regression begins. *Research:* Tilapia culture and biology.
1970s	*Development:* Regression continues; second wave of development begins. *Research:* Tilapia culture systems, pond culture systems, *Clarias,* "other" species biology
1980s	*Development:* The "second wave" continues in Côte d'Ivoire and Kenya. Small- and large-scale private-sector farming starts in Côte d'Ivoire, Egypt, Kenya, Nigeria, and Zambia; shellfish farming is initiated in Tunisia, South Africa, Morocco, Senegal, Zimbabwe, Malawi, Mauritius, and Reunion. A serious crisis of confidence occurs in subsistence-scale aquaculture, where 90% of development assistance is concentrated. *Research:* Tilapia culture systems, *Clarias,* oysters, "other" species biology, surveys, economic/commercial aspects, subsistence-level integrated aquaculture-agriculture
1990s	*Development:* Commercial development and diversification continues in many countries; production gradually mounts. Directions for small-scale aquaculture and contributions to food security of the poor are reexamined. Many countries and regional bodies (e.g., Southern African Development Community) start development plans for aquaculture. Private sector/governmental and nongovernmental organization roles are reexamined. Sustainability, natural resource management (including biodiversity), and climate considerations begin to impinge on the sector. *Research:* broader range of biological, socioeconomic, interdisciplinary, and intersectoral research

Source: Adapted from Powles (1987) and extended to the present period, with especial reference to Huisman (1990), New (1991), Satia (1991), Lazard et al. (1991), and Harrison et al. (1994).

In some African countries, a group of enterprises exists in a stage intermediate between rural and commercial development. These small-scale commercial systems are often the target group for development assistance projects because they represent the greatest hope for the transition to full commercial development (Satia, 1991; Brummett and Noble, 1995). These systems purchase some of their inputs but maintain their connections to the local communities. Failure by many of these enterprises to maintain this balance and thereby provide the predicted growth in aquaculture is widespread and has been part of the cause of widespread disillusionment with African fish farming (see, e.g., Lazard et al., 1991; Harrison et al., 1994; Huisman, 1990).

While commercial systems have not blossomed as expected, evidence exists that aquaculture is expanding among small farmers across the continent (Nathanael and Moehl, 1989; Molnar et al., 1991; van den Berg, 1994; Campbell, 1995; Murnyak and Mafwenga, 1995; Ngenda, 1995; Scholz and Chimatiro, 1995; ALCOM, 1994). This expansion now seems to be more demand- than technology-driven.

In Africa, experts are still divided on the question of the role of small-scale pond aquaculture in development. Since development efforts in Africa are dominated by help to small-scale farmers—more than 300 assistance projects were initiated from the early 1970s to the early 1990s, 90% of these in small-scale development—their lack of performance has led to a crisis of confidence and what Williams and Brummett (in press) call the "Great Small Scale African Pond Debate"—a debate over where assistance is best targeted and how its impacts should be measured.

The different views in this debate were somewhat clarified by Harrison et al.'s (1994) review. These authors pointed out that policy makers face the choice of targeting the resource poorest for food-security outcomes or focusing on technological developments that may be viable in the long term but which have little short-term impact on food security. Harrison and colleagues propose an overall aquacultural development strategy that hinges on planners first clarifying their objectives. These objectives may have to recognize that those most likely to benefit from fish farming will not be the worst off. In addition, food security outcomes are not well measured by the quantity of fish produced.

The relatively better-off farmers are usually the adopters of new practices (Harrison et al., 1994). In addition, Huisman (1986) pointed out that aquaculture is still a novel farming activity in much of Africa and therefore inherently risky for low-income people.

The novelty of the technology in Africa also suggests the need for a much greater research investment. To date, the share of worldwide research into African aquaculture mirrors the production level. A search of *Aquatic Sciences and Fisheries Abstracts* shows that about 1% of aquacultural papers (173 of 16,176) between 1978 and 1995 were on African aquaculture. The proportion rose marginally from 0.9% in 1987 to 1.15% in 1988 to 1995. Of the African papers, however, 38% were from South Africa.

Future Prospects. Katz (1995) has given Africa the lowest rating of any region with respect to commercial aquacultural potential, based on policy, physical, technical, production, and marketing attributes. Given the poor economic prospects for much of Africa at present, population growth will be the main force driving the demand for fish, compared with more balanced combinations of economic growth and population in other regions (Westlund, 1995). Africa exports and imports less fish than any other continent so markets are largely domestic. The difficult structural adjustments of the 1980s and 1990s have negatively affected all food production sectors. However, Delgado (1995) points out the relative inelasticity of demand for cheap fish in many sub-Saharan African markets. Together with population growth and urbanization, this inelasticity creates good outlets for smaller-scale producers who can efficiently grow low-value species.

As shown in other regions, national planning has helped pick up the pace, at the least, of commercial activities in countries around the Mediterranean. Ben Yami (1995) have reported that the governments of Egypt, Tunisia, and Algeria are promoting and supporting aquaculture through policies, services, and infrastructure and that this is paying off with increased production.

Rural development planning that incorporates subsistence aquaculture for household food security and environmental objectives will be more widespread over the next decade in inland Africa, spurred on by the general move for rural, participatory development as a tool for alleviating poverty. This development will have to be fully integrated with other rural sectors, assisted by extension and research services, and supported by schemes for affordable financing and supply of inputs such as fingerlings. Purchased inputs must be minimized, however, if the farms are to be viable. Inevitably, pond production will be low so that only small quantities of fish will be produced. Many of these fish will be eaten directly by the farmers' households or bartered at the pond side, thus losing little from postharvest handling.

ALCOM (1994), an interregional aquaculture development program, estimated that five southern African development community countries (Tanzania, Mozambique, Angola, Malawi, and Zambia) should be able to produce 250,000 metric tons of fish from inland ponds, compared with the present 5000 metric tons produced. Diffusion of the small-scale aquacultural technology is occurring in these countries where 25,000 ponds are now estimated to be farmed. ALCOM (1994) recommended assistance and research to increase the productivity of these ponds with a view to increasing the farmers' incomes.

There is a growing consensus that government roles need to be redefined (Lazard et al., 1991) and will be anyway since government resources are, at best, not increasing (ALCOM, 1994). The appropriate roles would include national planning and legislation, infrastructure support of appropriate kinds (e.g., roads to market rather than government-run hatcheries), research, extension, and information services.

Extension is a weak area, and expert opinions have been divided as to whether aquacultural training and technology transfer should rest with fisheries or agriculture departments. Most remain in fisheries departments, which have little rural extension capability and little experience of rural development.

As an African case study, ICLARM and GTZ (1991) examined the detailed context of integrated aquaculture-agriculture systems in Malawi. Malawi has a strong dependence on fish protein due to the capture fisheries of the nearby Great Lakes. The report concluded that aquaculture is not a panacea for alleviating poverty, food supply, and access problems. It is rather a complementary and supplementary technology for producing food. Many constraints to development exist, and these cannot be understood except within the full national and local contexts.

The constraints range from poverty (e.g., in Malawi, the average small farm is only 1.2 ha, nearly two-thirds of these farmers sell no food but consume it all themselves or barter it, and 55% of farmers are unable to even grow amounts sufficient to feed their families—Brummett and Noble, 1995), costs, labor (Christensen, 1995), and lack of expertise in digging ponds. Social constraints include local food habits, mistrust of government extension, and land tenure and ownership arrangements. (For example, the Malawian matrilineal system deters men from digging ponds when the land reverts to the wife's family on her death or after divorce—Brummett and Noble, 1955). Biotechnical constraints also limit the potential: nitrogen is limited, water is not readily obtainable on most farms, and feed is limited. Average fish productivity is low in ponds, as is most African agricultural productivity, which is also limited by low levels of inputs (fertilizers, trace elements, and water).

However, the ICLARM/GTZ report still concluded that aquaculture in Malawi could succeed technically and economically at different scales of operation, from the small-scale farmer integrating a small pond into a mixed farming system up to large-scale stocking of dams and commercial-scale operations operated by the government and by estate owners, especially sugar and tea estates.

Stocking of small water bodies, coastal lakes, dams, reservoirs, and even some of the Great Lakes is thought to offer considerable potential for increased African fish production. Coates (1995) has estimated that, for sub-Saharan Africa alone, the optimum potential yield from small water bodies is of the order of 1 MMT/yr, but that even less optimistic outcomes would provide substantial benefits. Environmental degradation, socioeconomic factors, and the need for good management regimes are the major constraints, there being few technical problems.

Coastal aquaculture is mainly commercial and carried out close to urban centers or for export (e.g., from the North African countries to Mediterranean Europe). Similar comments apply to the prospects for coastal aquaculture in Africa as for Asia and the island nations.

Latin America. Aquaculture production increased by 15% per year in the Latin American and Caribbean region between 1984 and 1992, but aquacultural products still represented only 1.86% of the per capita fish supply in 1992 (FAO, 1995a). Ecuador and Chile dominated production (36% and 21%, respectively), and Mexico, Brazil, Columbia, and Cuba each produced more than 20,000 metric tons. In many other countries, however, expansion is occurring rapidly.

Trade flows are strong for most of Latin America's fish production, and the FAO (1995b) expects that export market demands will dominate future developments in Latin American aquaculture, as they have for shrimp in Ecuador and salmon for Chile. One such example is the efforts in Ecuador to diversify shrimp production systems to grow flounder and Pacific yellowtail for export markets (Benetti et al., 1995). These export-driven developments are keeping attention away from the small-scale and subsistence production and therefore from production of affordable fish by and for low-income people. A major change in approach would be required in Latin America for aquaculture to have a major food security contribution.

Some rural development is happening through small-scale aquaculture projects. Castillo et al. (1992) and Popma et al. (1995) report on one such successful project in Guatemala. Boll and Lanzer (1995) have shown that low-intensity fish production is apparently not economically sustainable on small farms in Santa Catarina Province, the second southernmost province of Brazil. Despite this, and as a result of a 20-year extension effort and good availability of on-farm inputs and fry of several fish species, 10% of 170,000 small and middle scale farmers in Santa Catarina are engaged in fish farming. Boll and Lanzer (1995) recommend development of markets and of new technologies to affordably improve productivity.

Much Latin American aquacultural development has relied on introduced species such as salmons, carps, catfishes, trouts, and tilapia. Attention is now being turned to cultivating native species such as tambaqui (*Colossoma macropomum*) and pacu (*Piaractus mesopotamicus*) in Brazil (see, e.g., Castagnolli, 1995).

Overall, Latin American aquaculture is not yet giving high priority to the needs of small-scale producers and low-income consumers.

Gender

A gender analysis of the distribution of benefits from aquacultural development is important, yet few studies are available. Among the food-insecure in rural areas, the transition to greater fish scarcity and the increasing pressures on scarce farming land places greater burdens on all the household members. Women usually bear the main burden, especially as the fishing and farming activities of the men may now bring in less food and money. Women may have to help augment the family income beyond their previous contributions. They may have to help more in the farming, fishing, and aquacultural activities or gain income through outside activities. Frequently, men and young people migrate to the cities or other districts to find work, creating more single-parent and female-headed households.

Another major factor affecting women's roles in aquaculture is due not to events in the sector but to changes in the world social and economic order. Women are gaining stronger roles in society, including greater participation in education, the economy, and in decision making from the household to the national level. Women in aquaculture have not been immune to the greater focus on their roles, although few would claim that progress, however defined, has been great.

In fisheries and aquacultural development as in other sectors, the call for more knowledge of women's roles has been heeded, and more studies and projects are targeted on women, although the numbers are still low. For example, prior to 1975 there was fewer than one article per year on women in fisheries in the research literature. From 1975 to 1979, there was one article per year on average. Throughout the 1980s, the numbers grew to about 17 articles per year. Globally, eight fisheries development assistance projects had components on women by 1989, compared with four in 1985. More recent data are not available.

Men and women tend to undertake different tasks in aquaculture, although the divisions of labor vary between cultures and are not fixed and absolute. Men dig the ponds and are usually responsible for stocking them with seed. Women tend the ponds and feed the fish, men harvest them, and women process and sell the fish.

Greater involvement of women in aquaculture can give them greater financial independence, often a greater say in household decision making, and more money going directly to the family's food. However, in many countries women already have a strong and often controlling position in family financial matters.

Women's roles in aquaculture are often overlooked. They should be given greater recognition and have a higher profile in the sector to promote these roles. Studies reported at a recent seminar on "Women in Fisheries in Indo-China" (Williams, 1996) showed that the position of men in the household was critical to the choices available to women and to their status. Cambodia and Vietnam have high proportions of female-headed households, many due to recent wars. Female-headed households are usually worse off than those with male heads.

Demands on women's time are heavy in most societies. Hviding (1993) has pointed out that women in the Solomon Islands are heavily occupied in gardening, domestic work, reef gleaning, gathering of firewood, and raising their children. This all leaves little time to take up new activities such as clam farming. In Bangladesh, however, women are prevented by religion from working in the rice fields and going to market. They therefore have available time to undertake new activities such as fish husbandry.

New rice-farming technologies such as broadcast planting and applying herbicides, have reduced women's rice-growing labor in Vietnam and freed them for diversified farm activities including aquaculture (Vo Thi Ngoc Ba and Tran Thi Thanh Hien, 1996).

In addition to the usual food and financial benefits women derive from aquaculture, women in Bangladesh report gaining aesthetic pleasure from fish culture (Gupta and Rab, 1994).

Opportunities and Constraints

Regional, country, and gender differences aside, common patterns of opportunities and constraints emerge from the above analyses of how aquaculture benefits are distributed.

Constraints. Aquaculture can make a valuable contribution to improving the lives of the rural poor, but its development remains constrained by many factors.

Limited Extension Services. To date, research and development in aquaculture still lag behind those in other sectors of the economy (Pullin et al., 1993). As such, extension services geared toward promotion of technologies for and adoption of aquaculture are wanting. Extension workers often lack without sufficient training and expertise on the subject. Likewise, government support for extension work is insufficient.

In the Philippines, many areas suitable for aquaculture, and even areas where aquaculture is prevalent, are hardly ever reached by extension services, because of poor infrastructure, lack of budgetary support, and even lack of extension workers.

Similarly, in a study of two representative *thanas* (districts) in Bangladesh, only 7% to 8% of operators of farmed water bodies admitted receiving some form of extension service (Ahmed, 1992). For these reasons, most farmers engaged in or interested in aquaculture rely heavily on their own limited perceptions and their neighbors' and friends' experiences, rather than on more structured and scientific methods of aquaculture in their practice. This often results in a general disregard of "culture techniques and management procedures that are compatible with long-term capability of land and water resources." Women rarely receive formal aquacultural training, even though they are very active in the sector (Anonymous, 1996). Low levels of literacy among rural folk also hamper the effectiveness of existing efforts.

Poverty. Another limiting factor to the development of aquaculture is the abject poverty of many farmers in developing countries. Characteristically, incomes are much lower in farm households than in urban areas, so cash is limited and expenditures are concentrated to meeting basic household needs first. Thus, because of the rising cost of aquacultural inputs (such as feeds, chemicals, and fuel), coupled with difficulties in obtaining formal credit in rural areas, lower levels of inputs are used to reduce production costs, but this consequently leads to lower levels of production (Gupta et al., 1992).

Poor Infrastructure. For medium- to large-scale aquaculture enterprises, the poor condition of rural infrastructure also significantly affects aquacultural development. Lack of or impassable farm-to-market roads, poor telecommunication facilities, and unreliable electricity and transportation services limit aquaculture production and trade. In the Mekong Delta region of Vietnam for example, many fish farmers have difficulty in obtaining fries and/or fingerlings because of the distance to suppliers in the region, and with limited transportation available, obtaining them becomes an ordeal.

Water. Access to water of suitable quality is critical to the success of aquaculture. Marine and estuarine water quality and coastal space are severely limited by economic development and other uses. Aquaculture may itself produce effluents affecting water use. Freshwater quality and its availability are of increasing global concern (Gleick, 1993). Demand by aquaculture for adequate water will increase, although this use is rarely considered in major policies. Efficient systems for handling water, including closed systems, will need to be developed for many species, especially where intensification of culture increases the risks of negative off-site impact.

Challenge and Opportunities. Given due consideration, constraints to the development of aquaculture can challenge and provide opportunities for farmers, traders, researchers, government officials, and others.

The farmers' lack of financial resources and need for accessible credit facilities create opportunities for lending institutions to establish themselves in rural areas.

However, lending institutions are often hesitant to do so because of the perceived risks involved in lending to farmers. While farmers face a variety of production-related risks, creditors also face the risk of debtors defaulting on their loans. Thus, for lending institutions to even consider establishing and investing in rural areas, risks should be minimized. The risks are further increased in countries with high currency fluctuations and inflation.

Inadequacies in extension services can promote the establishment of cooperatives for fish farmers. Nongovernmental organizations often play a critical role in helping cooperatives get established. Once cooperatives are established, it would be easier and probably more effective for extension workers to disseminate information and training to well-organized groups than to individuals. Credit institutions likewise are more receptive to extending loans for bigger projects through cooperatives than for individual farmers because loans to cooperatives are easier to manage and monitor; hence the risk of default is lower. In this manner, the much-needed capital becomes more accessible to the fishers.

By integrating aquaculture into existing farming systems, costs can be minimized. It is imperative that aquaculture be considered as an integral part of the farming system for it to be viable and to get the most out of it. Expensive inputs such as chemicals, fertilizers, feeds, and even fuel and electricity should encourage farmers to try and make use of available resources within their farms. With proper planning and management, byproducts of other agricultural enterprises such as animal wastes can be used as biogas or as feeds and fertilizers for aquaculture, thereby saving on purchased inputs. Also, better farm management practices protect the environment and natural resources, which results in a more sustainable source of income and food (Lightfoot et al., 1993).

Many farm households, such as those in Bangladesh, have excess labor (Ahmed et al., 1995). The labor-intensive nature of constructing ponds and/or enclosing individual plots should serve as alternative employment opportunities for such surplus family labor. Christensen (1995), however, has pointed out that the returns on their labor are often low. Increasing pond productivity for higher returns therefore remains a challenge.

Governments should also be challenged to improve existing infrastructure facilities, or if nonexistent, to provide adequate infrastructure and telecommunication facilities. Doing so would help not only aquacultural development but rural development in general. With better infrastructure and telecommunications, more economic activities in rural areas can be generated, thereby contributing to the improvement of the quality of life in rural areas. Rural-to-urban migration, which has been recognized as a major problem in many developing countries, may also be discouraged.

There is much potential for higher foreign earnings from aquaculture (Pomeroy, 1992; New Zealand Trade Development Board, 1989), but lack of coherent and enforceable policies on international cooperation curb the development of aquaculture as a major foreign-exchange earning sector in developing economies. Governments must learn to take advantage of increasing world demand for aquacultural products and take appropriate steps to support aquaculture at all levels of production. A better accounting of the full costs of aquaculture will help to ensure that export returns are not gained at the expense of the local environment and livelihood.

Adoption Pathways

As an emerging technology, aquaculture has the potential to attract many new entrants and for existing practitioners to improve their production. In agriculture, extension services have been instrumental in introducing new methods and crops. In addition to the limitations discussed above, agricultural extension services are usually not technically competent in aquaculture. They are under challenge in many countries due to governmental financial stringency and to their perceived biases toward richer and male farmers and toward modern and exotic forms of farming (see, e.g., Chambers, 1983). Nongovernmental agencies are now playing a bigger role in technical extension in many countries.

Nevertheless, even extension programs involving government extension agents using well-tested methods developed by national research institutes can have major impacts on farmers' levels of knowledge and on their production of fish. In two districts of Bangladesh, fish production rose from the benchmark level of 618 kg/ha to 2728 kg/ha (Ahmed et al., 1995). The key factor influencing adoption before extension activities often was knowing about the technology. The intensity or degree of adoption was dependent on the size and previous history of culture of the water bodies used. Demaine and Turongruang (1996) have found that distance extension methods (e.g., radio programs, brochures, and videos) are effective in northern Thailand in reaching many more households than face-to-face services, especially because the extension materials and aquacultural techniques being promoted are well tested. However, better-off households are most likely to benefit.

The gap between existing knowledge and that required for a new complex technology such as integrated aquaculture-agriculture may be systematically narrowed as shown by a long-term (1982 to 1989) study in Guatemala. Castillo et al. (1992) have described the efforts to choose and train expatriate and local extension workers, and to introduce simple pond management and fish culture technology first and later progress to full integration of fish farming with small-animal production. The staged progression is seen as a major factor in the long-term success of integrated aquaculture-agriculture, especially where farmers have no tradition of culturing fish.

Causes of poor adoption of aquaculture include that it has been driven by available external technologies and that its promoters do not understand sufficiently the local conditions. In these cases, adoption requires more than promoting well-tested technologies. It needs simultaneous research, development, and adaptation of new technology.

Across all sectors, poor results in development projects have given rise to concepts of participatory development, a form of assistance with the dual aims of (1) empowering people by giving them a say in changes affecting their lives and (2) improving implementation efficiency by involving the target beneficiaries with their wealth of knowledge of local conditions (Mikkelsen, 1995). Participatory research has arisen from similar concerns over the poor adoption record for some research.

For aquacultural development, participatory extension and research approaches appear relevant where fish culture is not well developed or is not traditionally used. In such cases, adoption will occur only when practices, well suited to local conditions and the target beneficiaries, are developed.

Brummett and Noble (1995) have developed a participatory research approach because research experience in Malawi had shown that small-scale African farming systems, though using little modern technology, are complex, diverse, and not amenable to direct extension of existing technologies. In these authors' scheme, participatory research is instrumental in developing and modifying technologies that should then be better suited to adoption by farmers. Side benefits are two-way: the direct adoption of the technology by the small number of participating farmers and learning by researchers of real-world conditions. Harrison et al.'s (1994) review of African aquacultural development highlights extension issues from the farmers' and extensionists' perspectives; it recommends that extension workers be trained in more participatory approaches, basic aquacultural practice, and the role of ponds in the farming system beyond that of a fish production unit. Participatory extension approaches need improved coordination across government agencies, the private sector, and nongovernmental organizations (ALCOM, 1994).

DISCUSSION AND CONCLUSIONS

The demand for fish is rising in most parts of the world, and aquaculture will be the chief but not only source of supply to bridge the supply-and-demand gap. Aquaculture can contribute positively to food security in the developing world, but its contribution is neither automatic nor without problems. Maximizing this contribution requires certain interventions, including appropriate policies and planning, and development and adoption of new technologies. Interventions should be targeted especially to promote an equitable distribution of benefits and also the long-term environmental sustainability of aquaculture enterprises.

Aquaculture is a relatively new technology and hence risky for new entrants. Adoption depends on access to know-how, capital, water, space, and inputs such as fish seed and feed. It competes for space and resources with many other rural and coastal enterprises, sometimes to the detriment of other uses and users.

Aquaculture's most immediate and direct contribution to food security will come from maximizing the participation of small-scale enterprises that are producing whatever has the greatest immediate benefit to food security, be it low-cost food fish or high-value nonfood exports.

Mass aquaculture production of affordable fish for rural and urban consumers also contributes to food security. In some cases, such as small-scale aquaculture integrated with agriculture, the impact of aquaculture on sustainable food security is often not measured by fish production alone but by a range of fish and other crop products, and environmental and cultural benefits.

REFERENCES

Ahmed, M. 1992. Status and potential of aquaculture in small waterbodies (ponds and ditches) in Bangladesh. ICLARM Tech. Rep. 37. 36 p.

Ahmed, M., M. A. Rab, and M. P. Bimbao. 1993. Household socioeconomics, resource use and fish marketing in two thanas of Bangladesh. ICLARM Tech. Rep. 40. 81 p.

Ahmed, M., M. A. Rab, and M. P. Bimbao. 1995. Aquaculture technology adoption in Kapasia Thana, Bangladesh: some preliminary results from farm record-keeping data. ICLARM Tech. Rep. 44. 43 p.

ALCOM. 1994. Aquaculture into the 21st century in southern Africa. ALCOM Report 15. Working Group on the Future of ALCOM. Aquaculture for Local Community Development Programme, Harare, Zimbabwe. 48 p.

Ali, B. A. 1992. Rice-fish farming development in Malaysia: past, present and future. pp. 69–76, In, C. R. dela Cruz, C. Lightfoot, B. A. Costa-Pierce, V. R. Carangal, and M. P. Bimbao (Eds.). Rice-fish research and development in Asia. ICLARM Conference Proceedings 24. Manila.

Anonymous. 1995. 'Red tide' victims get food aid. Manila Bull., August 4, p. 32.

Anonymous. 1996. Draft summary. Seminar on Women in Fisheries in Indo-China Countries, March 6–8, Phnom Penh.

Auty, R. M. 1995. Patterns of development: resources, policy and economic growth. Edward Arnold, London.

Bailly, D., and W. Malistyani. 1996. Socio-economics of shrimp farming in Indonesia: a survey in Jepara and Pati Regencies, Central Java. pp. 25–26, In: R. LeRoy Creswell (Ed.). Book of abstracts: World Aquaculture '96. World Aquaculture Society, Baton Rouge, La.

Bell, J. 1995. Sustainable island aquaculture. pp. 12–13, In: J. Corbin (Ed.). Sustainable Aquaculture '95, June 11–14. Honolulu, Hawaii. Post-conference collection of meeting highlights. PACON 95

Ben Yami, M. 1995. Demand and supply of fish and fish products in north Africa: perspectives and implications for food security. pp. 75–92, In: Demand and supply of fish and fish products in selected areas of the world: perspectives and implications for food security. Paper presented at the International Conference on the Sustainable Contribution of Fisheries to Food Security, December 4–9, Kyoto. KC/FI/95/TECH/10.

Benetti, D. D., C. A. Acosta, and J. C. Ayala. 1995. Cage and pond aquaculture of marine finfish in Ecuador. World Aquacult, 26(4): 7–13.

Boll, M., and E. Lanzer. 1995. Exploratory bioeconomic study of fish polyculture in low intensity product systems in Santa Catarina, Brazil. pp. 16–23, In: Proceedings of the PACON Conference on Sustainable Aquaculture 95, June 11–14, Honolulu. PACON International, Hawaii Chapter.

Breth, S. A. (Ed.). 1996. Integration of sustainable agriculture and rural development issues in agricultural policy. Proceedings of the FAO/Winrock International Workshop on Integration of SARD Issues in Agricultural Policy, May 22–24, Rome. Winrock International, Morrilton, Arkansas.

Brummett, R. E. 1994. How can research best serve the needs of aquaculture in sub-Saharan Africa? Naga, ICLARM Q., 17(3): 15–17.

Brummett, R. E. 1995. The context of smallholding integrated aquaculture in Malawi: a case study for sub-Saharan Africa. Naga, ICLARM Q., 18(4): 8–10.

Brummett, R. E., and R. Noble. 1995. Aquaculture for African smallholders. ICLARM Tech. Rep. 46. Manila. 69 p.

Cai, R., D. Ni, and J. Wang. 1995. Rice-fish culture in China: the past, present, and future. pp. 3–14, In: K. T. MacKay (Ed.). Rice-fish culture in China. International Development Research Centre, Ottawa.

Campbell, D. 1995. The impact of the field day extension approach on the development of fish farming in selected areas in western Kenya. TCP/KEN/4551 (T) Field Document 1. Food and Agriculture Organization of the United Nations, Kisumu, Kenya.

Castagnolli, N. 1995. Status of aquaculture in Brazil. World Aquacult, 26(4): 35–39.

Castillo, S., T. J. Popma, R. P. Phelps, L. U. Hatch, and T. R. Hanson. 1992. Family-scale fish farming in Guatemala: an example of sustainable aquacultural development through national and international collaboration. Res. Dev. Ser. Int. Cent. Aquacult. 37. Auburn. 34 p.

Castro, E. R. 1995. So with rice, sugar, flour . . . : a drastic decline in 'bangus' fry supply. Manila Bull., August 18 pp. 13–14.

Chambers, R. 1983. Rural development: putting the last first. Longman Scientific & Technical, Essex, England.

Chong, K.-C., M. S. Lizarondo, Z. S. dela Cruz, C. V. Guerrero, and I. R. Smith. 1984. Milkfish production dualism in the Philippines: a multidisciplinary perspective on continuous low yields and constraints to aquaculture development. ICLARM Tech. Rep. 15. Manila. 70 p.

Christensen, M. S. 1995. Small-scale aquaculture in Africa—does it have a future? World Aquacult., 26(2): 30–32.

Coates, D. 1995. Inland capture fisheries and enhancement: status, constraints and prospects for food security. Paper presented at the International Conference on Sustainable Contribution of Fisheries to Food Security, December 4–9, Kyoto. KC/FI/95/TECH/3.

Coche, A. G. 1967. Fish culture in ricefields. A worldwide synthesis. Hydrobiologia, 30: 1–44.

Commission on Global Governance. 1995. Our global neighbourhood: the report of the Commission on Global Governance. Oxford University Press, New York.

Conway, G. R., and E. B. Barbier. 1990. After the green revolution: sustainable agriculture for development. 105 pp. Earthscan Publications Ltd., London.

Csavas, I. 1994. Aquaculture development planning in Vietnam. Paper presented at the National Workshop on Environment and Aquaculture Development, May 17–21, Haiphong, Vietnam.

David, C. C., and K. Otsuka (Eds.). 1994. Modern rice technology and income distribution in Asia. Lynne Rienner Publishers, Boulder, Colo.

Deelstra, H. A., H. Nuliens, and F. Adams. 1994. Nutritive value of fishes of Lake Tanganyika. II. Mineral composition. Afr. J. Trop. Hydrobiol. Fish., 5(1): 1–7.

Dela Cruz, C. R., C. Lightfoot, B. A. Costa-Pierce, V. R Carangal, and M. P. Bimbao (Eds.). 1992. Rice-fish research and development in Asia. ICLARM Conference Proceedings 24. Manila. 457 p.

Delgado, C. L. 1995. Fish consumption in sub-Saharan Africa. Paper presented at the International Conference on the Sustainable Contribution of Fisheries to Food Security, December 4–9, Kyoto.

Demaine, H., and D. Turongruang. 1996. Distance extension for aquaculture development in northeast Thailand. p. 103, In: R. LeRoy Creswell (Ed.). Book of abstracts: World Aquaculture '96. World Aquaculture Society, Baton Rouge, La.

De Silva, S. S. 1995. CGIAR aquatic research priorities revisited: a case for a higher priority for reservoir-lake system research. Naga, ICLARM Q., 18(3): 12–16.

Drèze, J., and A. Sen. 1989. Hunger and public action. Clarendon Press, Oxford, N.Y.

Edwards, P. 1996. Wastewater-fed aquaculture systems: status and prospects. Naga, ICLARM Q., 19(1): 33–35.

Edwards, P., and D. C. Little. 1995. Integrated crop/fish/livestock improvements in SE Asia. Paper presented at The Consultative Process to Develop ILRI's Global Agenda for Livestock Research, Consultation for the Southeast Asian Region, May 10–13, Los Baños, Laguna, Philippines. 18 p.

FAO. 1993. The state of food and agriculture 1993. FAO Agriculture Series No. 26. Food and Agriculture Organization of the United Nations, Rome.

FAO. 1994. The state of food and agriculture 1994. FAO Agriculture Series No. 27. Food and Agriculture Organization of the United Nations, Rome. 357 p.

FAO. 1995a. Aquaculture production statistics 1984–1993. FAO Fish. Circ. 815, Rev. 7. Food and Agriculture Organization of the United Nations, Rome, 186 p.

FAO. 1995b. Review of the state of world fishery resources: aquaculture. FAO Fish. Circ. 886. Food and Agriculture Organization of the United Nations, Rome, 127 p.

Gervis, M. H., and N. A. Sims. 1992. The biology and culture of pearl oysters (Bivalvia: Pteriidae). ICLARM Stud. Rev. 21. Manila. 49 p.

Gleick, P. H. (Ed.). 1993. Water in crisis: a guide to the world's fresh water resources. Oxford University Press, New York.

Gupta, M. V., and M. A. Rab. 1994. Adoption and economics of silver barb (*Puntius gonionotus*) culture in seasonal waters in Bangladesh. ICLARM Tech. Rep. 41. Manila 39 p.

Gupta, M. V., M. Ahmed, M. P. Bimbao, and C. Lightfoot. 1992. Socioeconomic impact and farmers' assessment of Nile tilapia (*Oreochromis niloticus*) culture in Bangladesh. ICLARM Tech. Rep. 35. 50 p.

Harrison, E., J. A. Stewart, R. L. Stirrat, and J. Muir. 1994. Fish farming in Africa: what's the catch? Summary report of ODA-supported Research Project "Aquaculture Development in Sub-Saharan Africa." University of Sussex, Falmer, Brighton, England. 51 p.

Hatta, A. M., and A. H. Purnomo. 1994. Economic seaweed resources and their management in eastern Indonesia. Naga, ICLARM Q., 17(2): 10–12.

Hettne, B. 1995. Development theory and the three worlds: towards an international political economy of development. 2nd ed. Longman Scientific and Technical, Essex, England.

Hiebert, M. 1995. Just compensation? Far East. Econ. Rev., 158(10): 14–15.

Huisman, E. A. 1986. Current status and rôle of aquaculture with special reference to the Africa region. pp. 11–22, In: E. A. Huisman (Ed.). Aquaculture research in the Africa region. Proceedings of the African Seminar on Aquaculture, October 7–11, 1985, Kisumu, Kenya. Pudoc, Wageningen, Netherlands.

Huisman, E. A. 1990. Aquacultural research as a tool in international assistance. Ambio, 19(8): 400–403.

Hviding, E. 1993. The rural context of giant clam mariculture in Solomon Islands: an anthropological study. ICLARM Tech. Rep. 39. Manila. 93 p.

ICLARM. 1995. From hunting to farming fish: background to press release.

ICLARM and GTZ. 1991. The context of small-scale integrated agriculture-aquaculture systems in Africa: a case study of Malawi. ICLARM Stud. Rev. 18. 302 p.

Jayaraman, N. 1995. Big fry vs. small fry: booming shrimp-farming business spawns protest. Far East. Econ. Rev., 158(2): 77–78.

Kapetsky, J. M. 1994. A strategic assessment of warm-water fish farming potential in Africa. CIFA Tech. Pap. 27. Food and Agriculture Organization of the United Nations, Rome. 67 p.

Katz, A. 1995. A study of national strategies for aquaculture development: commonalities in national aquaculture successes and failures. The results of an International Comparative Study of National Aquaculture Development carried out for the U.S. Congress, Office of Technology Assessment, 1994–1995. pp. 14–19, In: Sustainable Aquaculture '95 Symposium, June 11–14, Honolulu. Post-conference collection of meeting highlights. PACON 95: 14–19.

Kennedy, E., and H. E. Bouis. 1993. Linkages between agriculture and nutrition: implications for policy and research. International Food Policy Research Institute, Washington, D.C.

Kent, G. 1995. Aquaculture and food security. pp. 226–232, In: Proceedings of the PACON Conference on Sustainable Aquaculture '95, June 11–14, Honolulu. PACON International, Hawaii Chapter, Honolulu.

King, H. R. 1993. Aquaculture development and environmental issues in Africa. pp. 116–124, In: R. S. V. Pullin, H. Rosenthal, and J. L. Maclean (Eds.). Environment and aquaculture in developing countries. ICLARM Conference Proceedings 31. Manila. 359 p.

King, H. R., and K. H. Ibrahim (Eds.). 1988. Village level aquaculture development in Africa. Proceedings of the Commonwealth Consultative Workshop on Village Level Aquaculture Development in Africa, February 14–20, 1985, Freetown, Sierra Leone. Commonwealth Secretariat, London. 170 p.

Kokic, P. N., L. Moon, J. Gooday, and R. L. Chambers. 1995. Estimating temporal farm income distributions using spatial smoothing techniques. Aust. J. Stat., 37(2): 129–143.

Kyoto. 1995. Declaration and plan of action on the sustainable contribution of fisheries to food security. Presented during the International Conference on the Sustainable Contribution of Fisheries to Food Security, December 4–9, Kyoto.

Lazard, J., Y. Lecomte, B. Stomal, and J.-Y. Weigel. 1991. Pisciculture en Afrique Subsaharienne: situations et projects dans des pays francophone. Ministère de la Cooperation et du Developpement, Paris. 155 p.

Le Thanh Luu, Nguyen Huy Dien, N. Innes-Taylor, and P. Edwards. 1995. Aquaculture in the mountains of the northern Lao PDR and northern Vietnam. Naga, ICLARM Q., 18(4): 20–22.

Li, S. 1995. Opportunity and crisis of sustainable development of aquaculture in China. pp. 243, In: Proceedings of the PACON Conference on Sustainable Aquaculture 95, June 11–14, Honolulu. PACON International, Hawaii Chapter, Honolulu.

Li, S., and S. Xu. 1995. Culture and capture of fish in Chinese reservoirs. Southbound, Penang, Malaysia. 128 p.

Lightfoot, C., B. A. Costa-Pierce, M. P. Bimbao, and C. R. dela Cruz. 1992. Introduction to rice-fish research and development in Asia. pp. 1–10, In: C. R. dela Cruz, C. Lightfoot, B. A. Costa-Pierce, V. R. Carangal, and M. P. Bimbao (Eds.). Rice-fish research and development in Asia. ICLARM Conference Proceedings 24. Manila. 457 p.

Lightfoot, C., M. P. Bimbao, J. P. T. Dalsgaard, and R. S. V. Pullin. 1993. Aquaculture and sustainability through integrated resources management. Outlook Agric., 22(3): 143–150.

Mahata, S. C., M. G. Hussain, and M. A. Mazid. 1995. Artificial spawning of pond reared mahseer, *Tor putitora* (Ham.) in Bangladesh. Fourth Asian Fisheries Forum: abstracts. Asian Fisheries Society, Manila, Philippines.

McManus, J. W. 1995. Coastal fisheries and mollusk and seaweed culture in Southeast Asia: integrated planning and precautions. pp. 13–22, In: T. U. Bagarinao and E. E. C. Flores (Eds.). Towards sustainable aquaculture in Southeast Asia and Japan. Proceedings of the Seminar-Workshop on Aquaculture Development in Southeast Asia, July 26–28, 1994, Iloilo City, Philippines. SEAFDEC Aquaculture Department, Iloilo, Philippines.

Mikkelsen, B. 1995. Methods for development work and research: a guide to practitioners. Sage Publications, New Delhi, India. 296 p.

Molnar, J. J., A. Rubagumya, and V. Adjavon. 1991. The sustainability of aquaculture as a farm enterprise in Rwanda. J. Appl. Aquacult. 1(2): 37–62.

Munro, J. L. 1995. The scope for sustainable island aquaculture. Abstract of paper presented at the Sustainable Aquaculture '95 Symposium, June 11–14, June 1995, Honolulu.

Murnyak, D., and G. A. Mafwenga. 1995. Extension methodology practiced in fish farming projects in Tanzania. Paper presented at the Aquaculture for Local Community Development Programme (ALCOM), Technical Consultation on Extension Methods for Smallholder Fish Farming in Southern Africa, November 20–24, Lilongwe, Malawi.

Nathanael, H., and J. F. Moehl, Jr. 1989. Rwanda national fish culture project. Res. Dev. Ser. Int. Cent. Aquacult. 34. Allburn. 19 p.

New, M. B. 1991. Turn of the millennium aquaculture: navigating troubled waters or riding the crest of the wave. World Aquacult., 22(3): 28–49.

Newkirk, G. 1993. Do aquaculture projects fail by design? World Aquacult., 24(3): 12–18.

New Zealand Trade Development Board. 1989. Directions in: foreign exchange earnings. The New Zealand Aquaculture Industry, N.Z. Trade Development Board. V. R. Ward, Government Printer, Wellington.

Ngenda, G. 1995. Aquaculture extension methods in eastern Province, Zambia. Paper presented at the Aquaculture for Local Community Development Programme (ALCOM), Technical Consultation on Extension Methods for Smallholder Fish Farming in Southern Africa, November 20–24, Lilongwe, Malawi.

Pagdilao, C. R., L. G. Villacorta, and M. A. L. C. Corpuz. 1993. Status of the seaweed industry in the Philippines. PCAMRD-DOST Primer (19). Los Banos, Laguna, Philippines. 11 p.

Pauly, D., and V. Christensen. 1995. Primary production required to sustain global fisheries. Nature, 374: 255–257.

Phillips, M. J. 1995. Shrimp culture and the environment. pp. 37–62, In: T. U. Bagarinao and E. E. C. Flores (Eds.). Towards sustainable aquaculture in Southeast Asia and Japan. SEAFDEC Aquaculture Department, Iloilo, Philippines.

Pinstrup-Andersen, P. 1994. World food trends and future food security. Food Policy Rep. International Food Policy Research Institute, Washington, D.C. 25 p.

Pomeroy, R. S. 1992. Aquaculture development: an alternative for small-scale fisherfolk in developing countries. pp. 73–86, In: R. B. Pollnac and P. Weeks (Eds.). Coastal aquaculture in developing countries: problems and perspectives. International Center for Marine Resource Development, University of Rhode Island, Kingston, R.I.

Popma, T. J., R. P. Pelps, S. Castillo, L. U. Hatch, T. R. Hanson, and B. Duncan. 1995. Family-scale fish farming in Guatemala: an example of sustainable aquaculture development through national and international collaboration. pp. 299, In: Proceedings of the PACON Conference on Sustainable Aquaculture 95, June 11–14, Honolulu. PACON International, Hawaii Chapter, Honolulu.

Powles, H. 1987. Introduction to the workshop: history and status of African aquaculture research. pp. 2–13, In: H. Powles (Ed.). Research priorities for African aquaculture: report of a workshop held in Dakar, Senegal, October 13–16, 1986. IDRC-MR 129e. International Development Research Centre, Ottawa, Canada.

Prein, M., J. K. Ofori, and C. Lightfoot (Eds.). 1966. Research for the future development of aquaculture in Ghana. Summary papers of the ICLARM/IAB Workshop, March 11–13, 1993. Institute of Aquatic Biology, CSIR, Accra. ICLARM Conference Proceedings 42. Manila. 94 p.

Pullin, R. S. V., H. Rosenthal, and J. L. Maclean (Eds.). 1993. Environment and aquaculture in developing countries. ICLARM Conference Proceedings 31. Manila. 359 p.

Ratafia, M. 1995. Aquaculture today: a worldwide status report. World Aquacult., 26(2): 18–24.

Rogers, J. 1990. What food is that? and how healthy is it? Welson, Sydney.

Satia, B. P. 1991. Why not Africa? Ceres, 23(5): 26–31.

Scholz, U., and S. Chimatiro. 1995. The promotion of small-scale aquaculture in the southern region of Malawi: a reflection of extension approaches and technology packages used by the Malawi-German Fisheries and Aquaculture Development Project (MAGFAD). Paper presented at the Aquaculture for Local Community Development Programme (ALCOM), Technical Consultation on Extension Methods for Smallholder Fish Farming in Southern Africa, November 20–24, Lilongwe, Malawi. 11 p.

Soemarwoto, O. 1990. Introduction. pp. 1–6, In: B. A. Costa-Pierce and O. Soemarwoto (Eds.). Reservoir fisheries and aquaculture development for resettlement in Indonesia. ICLARM Tech. Rep. 23. Manila. 378 p.

Speth, J. G. 1993. Towards sustainable food security. Sir John Crawford Memorial Lecture, International Centers Week, October 25. Consultative Group on International Agricultural Research, Washington, D.C.

Thorne-Miller, B., and B. Jaildagian. 1995. A framework for sustainable marine aquaculture: can it work in the Caribbean region? Paper presented at the Sustainable Aquaculture 95 Symposium, June 11–14, Honolulu.

Tyler, P. E. 1995. China's worried peasants unimpressed by year of the farmer. International Herald Tribune, April 11, p. 4.

van den Berg, F. 1994. Privatization of fingerling production and extension: a new approach for aquaculture development in Madagascar. pp. 32–34, In: R. E. Brummett (Ed.). Aquaculture policy options for integrated resource management in sub-Saharan Africa. ICLARM Conference Proceedings 46. Manila. 38 p.

Villanueva, M. A. 1994. Fishponds, prawn farms freed from land reform. Manila Standard, November 19, p. 2.

Vo Thi Ngoc Ba and Tran Thi Thanh Hien. 1996. Preliminary study on the contribution of women in the income structure from agricultural production at Songhau farm, Mekong Delta of Vietnam. Paper presented at the Seminar on Women in Fisheries in Indo-China Countries, March 6–8, Phnom Penh, Cambodia. 5 p.

Westlund, L. 1995. Apparent historical consumption and future demand for fish and fishery products: exploratory calculations. Paper presented at the International Conference on the Sustainable Contribution of Fisheries to Food Security, December 4–9, Kyoto. KC/FI/95/TECH/8. 55 p.

Williams, M. J. 1996. Women in fisheries in Indo-China countries. Paper presented at the Seminar on Women in Fisheries in Indo-China Countries, March 6–8, Phnom Penh, Cambodia. 6 p.

Williams, M. J., and R. E. Brummett. In press. Aquaculture in African rural and economic development.

3 Offshore Mariculture*

ROBERT R. STICKNEY

INTRODUCTION

Multiple-use conflicts and environmental concerns associated with the nearshore marine waters of much of the United States and, increasingly, many other nations, have imposed significant limitations on the development of mariculture in the coastal zone. Objection to the culture of salmon in net pens within Puget Sound, Washington, presents a good example of the issues that have been raised. Those issues include visual pollution; potential impact on nontarget bacteria from the use of antibiotics in fish feed; infection of wild fish with diseases carried by cultured fish; noise and odor pollution; pollution of the water column and destruction of the benthic community from waste feed and fecal deposition; interference with navigation; and removal of access to traditional commercial fishing, sportfishing, and recreational areas. The issues have hampered aquacultural development in the Puget Sound region, though proper management, including proper siting of facilities, can allow salmon net-pen culture to proceed in an environmentally responsible manner (Parametrix, 1990). Objections by competing user groups may never be fully overcome.

One of the most frequently mentioned options for expansion of the mariculture industry is to move offshore into more open waters. Within 4.8 km of the shoreline the states have jurisdiction over maricultural development. Further offshore, from 4.8 to 320 km, in the region known as the Exclusive Economic Zone (EEZ), mariculture will come under the jurisdiction of the federal government. Extension of mariculture to the open ocean has been discussed for a number of years, though truly

*This chapter was developed from a report submitted by the author to the Office of Technology Assessment (OTA), U.S. Congress, entitled "Offshore Aquaculture: Technology and Policy Issues." That report was prepared with input from members of the OTA Offshore Aquaculture Committee. The committee members also provided useful criticism and additional input on the OTA report draft. Members of that committee were Bob Blumberg, John Forster, Harlyn Halvorson, James Harding, Conrad Mahnken, James McVey, Russell Miget, Granvil Treece, and Robert Wilder. The assistance of OTA staff members Alison Hess and Robin White in all phases of committee activity is gratefully acknowledged.

Sustainable Aquaculture, Edited by John E. Bardach
ISBN 0-471-14829-6

open-ocean commercial fish or shellfish facilities (those located in the EEZ) were not in evidence in any nation at least through 1993 (Sveälv, 1991a; McCoy, 1993a). Commercial systems had been established at least 2 km offshore by that year (McCoy, 1993b), and research had been conducted 80 km offshore in the Gulf of Mexico in 70 m of water (Linfoot et al., 1990; Loverich, 1991a).

Commercial mariculture is currently being proposed or, in at least one or two cases, is underway in offshore areas. Research with submerged cages 11 km offshore from Israel in the Mediterranean Sea beginning in 1992 has been sufficiently successful that a commercial facility is being developed (Anonymous, 1995b). In 1994, a firm expressed its intention to establish a salmon-rearing facility 85 km off the coast of Massachusetts in the Atlantic Ocean (Plante and Jones, 1994). Gilthead sea bream have been under culture in the Red Sea 20 km off the coast of Israel since early 1994 (Chamish, 1995). Various systems designed for offshore use continue to be tested in exposed nearshore waters.

As discussed by Fridley (10), the development of offshore mariculture facilities "could alleviate many of the institutional, regulatory, and environmental problems associated [with] coastal marine mariculture." Moving mariculture to offshore locations would avoid impacting the more environmentally sensitive nearshore areas of the coastal zone (National Research Council, 1992), and there might even be some benefits in terms of improved product quality (Sveälv, 1988, 1989). Movement of facilities to offshore sites would require some basic changes, not only in the types of materials used for the manufacture of culture facilities, but also in the types of systems employed, the techniques used in conjunction with those facilities, the support industries, and to some extent, the species cultured. All those considerations and the attitude with which mariculture is approached would have to be oceanic in concept (Loverich, 1991b).

The scope of offshore mariculture has yet to be delineated, but it could certainly be quite broad, limited only perhaps by economics. That is, if there is sufficient profit potential from a species or group of species being contemplated for culture to justify the construction and operating costs of the proposed facilities, venture capital would become available and an industry would be developed. How large the industry becomes would depend on demand for the products. Because overhead costs would be much higher for the average offshore mariculturist than for those operating in protected coastal waters, luxury species would undoubtedly be given highest priority. Atlantic salmon (*Salmo salar*) are a clear choice for cold-water regions. Pacific salmon such as chinook (*Oncorhynchus tshawytscha*) and coho (*O. kisutch*), Atlantic halibut (*Hippoglossus hippoglossus*), and Pacific halibut (*H. stenolepis*) are additional possibilities. In warm-water regions, gilthead sea bream (*Sparus aurata*), red sea bream (*Pagrus major*), seabass (*Dicentrarchus labrax*), yellowtail (*Seriola quinqueradiata*), and dolphin (*Coryphaena hippurus*, also known as mahi-mahi) hold considerable potential, and in some cases are already being reared in exposed areas.

Polyculture may offer significant opportunities for the development of economically viable production facilities for species that would be uneconomical grown in monoculture in offshore areas. Mollusks and kelp could be reared in conjunction with fish. One can envision a polyculture system in which kelp are grown to provide food for abalone (a high-value crop) or harvested for chemicals such as agar and carageenan. Other species of mollusks and perhaps fish could also be grown in a properly designed facility. Nutrients released from animal production facilities could be converted to plant biomass in polyculture facilities. Many offshore areas are undoubtedly nutrient-limited and might not be able to support significant levels of seaweed production unless there is a nutrient input source. Thus, developing kelp culture facilities in association with net pens may be an option that would provide for some recycling of nutrients released from the net pens and incorporated into seaweed (Petrell and Mazhari, 1992).

In the future, a wide variety of marine organisms may be reared as sources of pharmaceuticals. Offshore maricultural systems may someday be developed exclusively for the production of species that are used in conjunction with such products.

Offshore mariculture presents a unique opportunity for sustainability. It has been demonstrated that net-pen fish culture can be practiced in coastal waters without damage to the environment if the facilities are properly sited (Parametrix, 1990). If such facilities are placed in offshore waters where circulation is excellent, and if the facilities are properly sized, no significant negative impact on local environmental conditions should occur. Thus, offshore mariculture represents a next logical step in the development of the industry. Culture facilities—such as those sited in offshore waters—which can be operated without negative impacts on the local environment represent, by definition, sustainable systems. While it is conceivable that a saturation point can be reached for offshore maricultural systems, the vast expanse of the world ocean suggests that the ultimate demand for protein from the sea will be reached, due perhaps to the ultimate maximization of the human population, long before all potential offshore mariculture sites are occupied. Careful attention to the density of offshore mariculture facilities within a given locale will ensure that the sustainability of the activity is maintained.

ADVANTAGES OF OFFSHORE MARICULTURE SYSTEMS

Coastal regions throughout much of the world are becoming increasingly polluted. Mariculture facilities located in nearshore waters can be subjected to such pollutants as detergents, sewage, food-processing wastes, thermal effluents, trace metals, pesticides, and oil spills (Beveridge, 1987). Significant pollution problems have impacted maricultural activities in Japan, the Philippines, and elsewhere. Developing nations are beginning to recognize that pollution is having a negative impact on their coastal aquaculture facilities, but the problem remains to be aggressively addressed (Pullin et al., 1993). Self-pollution—that is, the pollution of water used in mariculture by the wastes from the culture operation or those adjacent to it—has become a problem in some locations.

Prohibitions on shellfish harvesting because of sewage contamination and toxic algal blooms are becoming increasingly common along all the coastlines of the United States and in other locations around the world. In some instances the causes of the problem have yet to be determined, but often, contamination of shellfish beds is a result of the activities of humankind. Pollution abatement programs and the strict controls that have been implemented in many developed nations make the coastal waters in those countries more suitable for mariculture than are the coastal areas of nations that have not attempted to curb pollution. Regardless, shellfish bed closures can be expected to occur virtually anywhere as a result of inadvertent releases of pollutants. Such releases can have significant deleterious effects not only on the natural environment, but also on coastal mariculture facilities. Such problems can be significantly reduced by moving mariculture offshore.

Aquaculture facilities can have a variety of impacts on natural environments, just as the environment in the immediate vicinity of such facilities can impact the species under culture. The environmental impacts of aquaculture are associated mainly with husbandry techniques, feeding, feed composition, and site selection (Parametrix, 1990; Ackefors, 1992, 1993). By locating mariculture facilities in deep, open-water areas, the dispersion of nutrients released from fish and shellfish farms will be enhanced, and benefits may actually accrue in terms of increasing local floral and faunal production levels. Sport and commercial fishing could be enhanced, not only because the structures associated with offshore mariculture facilities will serve as fish-attracting devices but also because of increased nutrient levels, which will promote increased overall local productivity. Whereas increased nutrient levels in coastal waters may be considered a negative impact, modest increases in offshore nutrient levels may actually be a benefit.

Proper siting of net-pen facilities will provide for the dispersion of wastes in a manner that will not lead to significant eutrophication (Parametrix, 1990). As mariculture facilities are moved offshore, overall environmental impact can be reduced relative to that produced by coastal facilities (Dahle, 1991b). Mathematical dispersion models are available that can be used to ensure that proper sites are selected (Bell and Barr, 1990). Gowen and Edwards (1990) pointed out that the capacity of offshore sites to accept waste from fish farms can be exceeded if the intensity of culture becomes excessive, though even in nearshore sites the inputs of phosphorus and nitrogen from aquaculture are very small compared with municipal, industrial, and agricultural and nonagricultural river inputs (Ackefors, 1993). Thus, some control on the number of facilities within a given area (and the total number of animals under culture) must be exerted to maintain waste levels within the assimilation capacity of the environment. Establishing large sites with relatively low biomass per unit area, and the use of widely distributed net pens or drifting systems that have single-point mooring, will help ensure the proper dispersal of released nutrients.

Offshore aquaculture may benefit from vertical integration wherein companies growing the products would also be involved in processing and marketing. Processing could be accomplished offshore or in land-based plants. Existing regulations for shipboard processing in the EEZ would undoubtedly also apply to offshore processing of aquacultural products. Much or all of the processing waste from aquatic animal production could be recycled back to the net pens as a constituent of the feed, thereby greatly reducing or eliminating the disposal problem.

In terms of water-quality fluctuations, offshore aquaculture sites will provide less variability than inshore locations. Many water-quality variables remain virtually constant throughout the year in offshore sites. Examples are salinity and the levels of most ions dissolved in seawater. Water-quality variables that do fluctuate significantly tend to undergo slow changes in deep offshore waters as compared with inshore shallow areas, and the range of fluctuation in offshore waters may be somewhat less than what is seen in the nearshore environment. Temperature, dissolved oxygen, nutrient levels, and turbidity are examples.

Noxious phytoplankton blooms have caused severe fish losses in the United States, Canada, Norway, and Japan. Such blooms may be toxic, but more commonly lead to mortality of cultured fish by mechanically clogging the gills. Such phytoplankton blooms are less likely to develop offshore than in coastal waters. In offshore areas where such blooms are an issue, submersible net pens could be used to place the fish below the algae. Such blooms tend to be concentrated in the near surface water, so net pens or cages would not necessarily have to be submerged more than a few meters.

From a research standpoint, offshore aquaculture sites could be used as locations for the collection of not only biological information, but also meteorological and oceanographic data (Wilder, 1992). Instrument arrays could easily be added to aquaculture facilities without interfering with aquatic organism production. Collection of meteorological and oceanographic information could be of value to the scientific community and the military, and would also provide information useful to aquaculturists. Oil and gas production platforms provide excellent opportunities for meteorological and oceanographic research and monitoring (Cooper et al., 1993; Dacey and Cooper, 1993; Dokken, 1993; Jeffress et al., 1993; Lewis, 1993; Roscigno and Kennicutt, 1993; Taylor et al., 1993; Wiebe et al., 1993), which could easily continue if a platform were to be converted to aquaculture production.

The maintenance of biodiversity is a subject of increasing interest among ecologists and policy makers. It can be argued that marine biodiversity could be aided by the use of offshore mariculture. Fishing and other human activities that have led to habitat alteration and depletion of certain stocks may be reducing the diversity of marine organisms in various regions both inshore and in the EEZ. By employing offshore maricultural systems to reproduce a variety of native organisms, some amelioration of the situation could be provided through the reintroduction or augmentation of species that have been impacted by human activities.

NONTECHNOLOGICAL CONSTRAINTS TO DEVELOPMENT OF OFFSHORE MARICULTURE

A number of constraints to offshore mariculture, other than the need for appropriate technology, have been identified. Some may be more perceptual than real, and others will require the development of governmental policies in order to be properly addressed. Many of the topics discussed here have also been addressed by DeVoe (Driessen, 1987).

The marine environment has been and continues to be used as a repository for various types of waste. The site selection and permitting process associated with development of offshore mariculture facilities should take into account past uses for any location being evaluated. With respect to contamination, that evaluation should focus primarily on what use had been made of the bottom as a repository for wastes. Examples of areas that should be avoided include those that have been used for the disposal of trash, toxic material, medical waste, radioactive substances, and munitions. Dredge material is often dumped offshore, creating localized turbidity increases, and in some cases toxicity within both the sediments and the water column. Dredge-material disposal areas would need to be evaluated on a case-by-case basis to determine their suitability as locations for offshore mariculture sites. It would not seem appropriate to establish a maricultural operation in the vicinity of a dredge-material disposal area that is active or that is subject to reactivation.

The reliability of offshore culture systems to function properly when personnel cannot be continuously on-site is another identified constraint (National Research Council 1992). Accelerated research and development of automated systems for offshore mariculture will be necessary if that constraint is to be effectively overcome. Having personnel on hand to make frequent inspections of various components for signs of structural fatigue and to ensure that the animals are performing properly will help operators avoid catastrophic losses. Malfunctions in automatic systems such as feeders and water-quality monitoring equipment can often be avoided or quickly remedied when technical staff are routinely present. However, it may be difficult or impossible to have personnel continuously present at offshore facilities. Facilities may involve a number of widely distributed structures serviced by roving personnel. Only fixed structures of relatively large sizes could provide accommodations suitable for personnel to remain on board for periods of days or weeks. During bad weather it may not be possible for personnel to visit unoccupied satellite sites at the normal frequency. Thus, automated systems will be a requirement. Feeders need to have sufficient capacity to continue operating for several days during inclement weather.

Some manufacturers now provide guarantees of up to 12 months on their marine net-pen systems (McCoy, 1993a). It can be assumed that such guarantees remain in force only as long as the systems are operated within a prescribed range of conditions, which may not include being located in highly exposed areas.

As discussed by Clay et al. (1981), the need to establish and protect the property rights of the private offshore mariculturist is paramount. Animals and plants that are under the care and protection of an offshore mariculturist and held within a leased or licensed site must be declared private property so they can be harvested and sold as necessary. As discussed by Eichenberg and Vestal (1992) in conjunction with the development of mariculture systems in territorial waters, private use or control of marine organisms is not a part of the well-established property rights system that exists with respect to traditional agriculture and other inland activities. However, if mariculturists are not given ownership of the organisms being reared, there will be little incentive to develop an offshore industry. This situation has arisen in some states in conjunction with the development of the freshwater aquaculture industry in instances where resident aquatic organisms were legally the property of the state. Those types of statutes have been changed in most instances to provide aquaculturists ownership and the right to freely harvest and market their animals. Ownership of fish in the marine EEZ is considered to be held by the public at large, where management is conducted by the federal government. Laws providing ownership of species farmed in the EEZ to the mariculturist would have to be promulgated.

Offshore mariculture facilities, whether within the jurisdiction of the state or federal government will be sited in public waters, and, while they may not be permitted in or immediately adjacent to established shipping lanes, such facilities could present hazards to navigation. Radar reflectors, foghorns, and proper lighting will be required to ensure that vessels will be able to detect the presence of mariculture facilities regardless of time of day and weather conditions. Required navigational aids may vary with the size of the mariculture facility and the type of mooring system used. Existing guidelines for marking oil and gas platforms and other offshore structures should be applicable to most offshore mariculture sites.

Appropriate offshore mariculture sites can be expected to be found, in many instances, in traditional commercial fishing areas, so development of culture facilities will undoubtedly be considered an impediment to the use of some types of fishing gear. In most cases, at least a limited amount of mariculture should be possible without affecting normal catch rates from commercial fishing grounds. For example, the presence of several thousand oil production platforms in the Gulf of Mexico, many of which can be found on traditional shrimp fishing grounds, provide an indication that different user groups can coexist. Trawlers not only effectively navigate around production platforms, they must also be aware of obstructions on the bottom, such as natural and artificial reefs, along with wrecks and other debris that can entangle and destroy nets. The addition of a mariculture component to existing production platforms would not significantly change the existing situation. Where new offshore mariculture facilities are established as independent operations over traditional fishing grounds, site size can be controlled to minimize the loss of access by fishermen.

It is unlikely that offshore mariculture will ever utilize more than a minute portion of the available EEZ, even as the industry matures. Relatively small numbers of sites, each only a few hectares in area and efficiently operated, can produce sufficient quantities of certain species to satisfy current and future demands. The requirement to produce high-value products will, in turn, mean that the market for species grown in the U.S. EEZ will be largely domestic. Export of products reared in offshore facilities will only be to countries with large numbers of people who are able to afford the premium prices required from fish and shellfish cultured in offshore facilities.

Establishment of offshore mariculture will undoubtedly be prohibited over areas of the EEZ currently restricted to military use only. Impedance of military surface shipping, in general, would be no different from that associated with commercial or private vessels, though special navigational aids may be required in areas where submarines operate.

Interactions between offshore mariculture facilities and the natural environment include at least the potential for undesirable genetic impacts from escaped hatchery fish interbreeding with wild fish. The possibility of disease transmission from cultured to wild stocks has also been mentioned as a negative factor (Donaldson, 1991; Fridley, 1991). In terms of disease transmission, cultured fish can also be at risk from wild stocks (Donaldson, 1991).

A considerable amount of controversy surrounds the use of genetically altered animals in mariculture. Transgenic species provide enormous potential, but the perceived risks associated with their use often receive more attention than the benefits that might accrue. Fish and shellfish could be genetically altered to be resistant to various diseases, demonstrate improved growth, and yield increased dress-out percentages. So-called antifreeze genes could be implanted to increase the temperature tolerance of warm-water species, and a variety of other improvements in performance could result. A major concern surrounds the escape of genetically altered fish into the surrounding environment where they might spawn with and alter the genetic structure of wild stocks. Various techniques, including chemical sterilization and polyploidy, could be used to produce animals for stocking that are unable to reproduce.

Many of the predators that have caused problems for nearshore and onshore mariculturists can be expected to plague offshore fish farmers as well. Included are a number of species of fishes, turtles, sea snakes, piscivorous (fish-eating) birds, marine mammals, and even squid (Beveridge, 1987). Major predators, including all marine mammals and many piscivorous birds, are protected under U.S. law, so lethal means of controlling them are either unavailable or severely restricted. Permits to eradicate small numbers of some predators can often be obtained, but the effort typically has little or no impact on total fish losses since the predators are typically present in far greater numbers than are allowed to be controlled under the permits issued. Predator nets around net-pen facilities have been successfully employed by nearshore fish farmers, and bird netting over the top of net pens is generally effective if properly maintained. Submerged net pens are not be susceptible to bird predation but may require predator nets.

The presence of any type of structure in the marine environment will attract various types of animals. Fish congregate under floating debris, in the vicinity of any protuberance in areas where the bottom is otherwise flat and featureless, and around such structures as gas and oil platforms and maricultural net pens. The prepared feed used in conjunction with the rearing of many species will also attract fish and other organisms. Although the increased density of wild fishes around a mariculture facility can provide enhanced recreational fishing opportunities for the angling public, it may be detrimental to the mariculturist. The mariculturist may have increased liability for injuries suffered by people fishing around an offshore facility.

Security has been a common problem for aquaculturists (Edwards, 1978; Secretan, 1980; Beveridge and Muir, 1982). While on the surface it might appear that poaching, equipment theft, and vandalism would be reduced at offshore mariculture sites, the fact that offshore facilities would be located in public waters makes them accessible and vulnerable. Submerged net pens would be less subject to poaching, but auxiliary apparatus such as floats, mooring lines, monitoring equipment, storage sheds, and feeding equipment located at the surface would be susceptible to vandalism. Video surveillance and the use of alarm systems may be effective deterrents. Severe penalties for theft and vandalism might also dissuade at least some people from participating in illegal activities.

Escape from mariculture facilities can and does occur, even when great care is taken to prevent that type of loss. When exotic species are being cultured, there is a chance that escapees will become established, perhaps to the detriment of one or more native species. It may therefore be prudent to avoid the culture of exotic species except for the use of sterile hybrids, triploids, or unisex stocks so that escapees would be unable to establish reproducing populations. (See Chapters 2 and 5.)

Price volatility can be an important constraint to all forms of mariculture. If prices paid to producers are subject to wide fluctuations, the viability of offshore mariculture operations could be threatened.

Lack of access to capital and high insurance premiums were cited as impediments to offshore mariculture development by Clay et al. (1981). Obtaining crop insurance for offshore mariculture facilities may not be impossible, but it will undoubtedly be expensive. Findlay (1990) has reported that underwriting profits for traditional inshore salmon net-pen facilities have not been sufficient and that significant weather-related losses have occurred. It is important that the insurer selected by an aquaculturist understand the difference between inshore and offshore sites (Smith, 1990). In the United Kingdom, Lloyd's Register has developed a classification system for fish farms. Classification is based on the design and manufacturing process used, site location, mooring design, and inspection of the actual installation (Barker, 1990).

APPROACHES TO OFFSHORE MARICULTURE

Working in offshore areas frequently requires vastly different approaches and facilities than those used in coastal or inland waters. Constant stress from waves and currents, with intermittent exposure to storms that greatly increase the stress levels, require the use of materials and engineering approaches that are much more expensive than those used in protected waters. Support equipment, such as feeding systems and the boats used to service mariculture facilities, must also be able to withstand the rigors of the offshore environment. Because of water depth, bottom culture is not generally a viable option for offshore mariculture. Various types of production systems have been envisioned, and several prototypes have been constructed and tested. Approaches include confinement of the cultured species in fixed or mobile units and consideration of unconfined culture. While most trials to date have been conducted at nearshore, though often highly exposed, sites where conditions similar to those that occur in the EEZ, there has been some offshore experimental activity, as for example the testing of various cage designs in the Gulf of Mexico and the Strait of Juan de Fuca (Loverich, 1991a; Loverich and Croker, 1993; Loverich and Swanson, 1993; Miget, 1995). Also, there has been some interest expressed in establishing commercial mariculture ventures off the eastern coast of the United States (Plante and Jones, 1994). Other nations in which interest in offshore mariculture has developed and studies are being conducted include Canada (Moffatt, 1991), Cyprus (Muir and Baird, 1993), France (Anonymous, 1995a), Hong Kong (Bingham, 1991), Israel (Anonymous, 1995b; Chamish, 1995), Italy (Lazzari and DiBitetto, 1993), Japan (Okamoto et al., 1993), Scotland (Götmalm et al., 1989), and Sweden (Sveälv, 1988). The major oceans are not the only sites being considered. Interest in establishing offshore mariculture systems in the Black and Caspian Seas has also been expressed (Muravjev et al., 1993).

Fixed and Mobile Systems

Fixed structures include permanent facilities anchored to the seabed from which culture chambers are suspended and moored. Net-pen systems can be comprised of either floating or submerged structures. Floating offshore maricultural structures may have single-point mooring, which allows them to swing with currents, or they may have multiple moorings to keep them in a fixed location. Fixed systems can be kept in place with an array of anchors and cables, or they may be attached to structures that are standing on the bottom (Miget, 1995). One manufacturer has been experimenting with floating net pens that feature spar buoys that help stabilize the culture chambers during storms (Loverich, 1991b; Loverich and Croker, 1993; Loverich and Swanson, 1993).

In addition to the grow-out of fish or shellfish to market size on fixed offshore facilities, a number of other possibilities exist. Offshore facilities might in some cases be used primarily or solely as hatcheries (Anonymous, 1995a) from which fish or shellfish juveniles would be distributed to other offshore, nearshore, or onshore grow-out facilities. Animals may be grown for sale to processors, or they may be released to enhance commercial, or more likely, sport fisheries. Offshore facilities could also be used to depurate (cleanse) shellfish that have been maricultured or commercially captured in contaminated areas.

Platforms located at or above the ocean surface, but standing on legs that reach to the seabed, can serve multiple purposes: as mooring sites for fish cages or net pens, as attachment points for longlines of mollusks, or as points for suspending nets to support seaweeds. Constructing such facilities only for mariculture may not be economically feasible, but combining mariculture with some other commercial activity could be profitable. One example involves the development of mariculture in conjunction with offshore oil and gas production platforms. Such platforms could be used for the hanging culture of such shellfish species as oysters, scallops, and mussels as well as for tethering net pens (National Research Council, 1992; Miget, 1995).

In some instances aquaculturists may be able to take advantage of existing offshore structures to which they could append their facilities. The best examples are oil and gas production platforms (Figure 3.1). There are about 4000 such platforms in the Gulf of Mexico at the present time (Driessen, 1987). While a platform might have a useful working life of up to 50 years, its functional life with

Figure 3.1. An oil and gas platform in the Gulf of Mexico. A prototype OceanSpar™ offshore net-pen system can be seen to the right of the platform.

respect to oil or gas production is normally measured in terms of a decade or two. Conversion from an oil and gas production mode to mariculture would represent an alternative use; the justification for this, after cessation of hydrocarbon production, remains to be legally defined since current regulations require the removal of platforms once they are no longer producing oil or gas.

The National Research Council (1985) has indicated that the total cost of platform removal in the Gulf of Mexico could exceed $1 billion by the year 2000 and could reach a total of $7.5 billion by 2020. If a production platform is located in 60 m of water or less, the cost for its removal may reach $4 million, while the cost to remove a platform located in water 60 to 120 m deep may be as high as $10 million.

Net pens, longlines, and rafts can be moored to, or suspended from oil and gas production platforms. There will undoubtedly be limitations on the amount of weight that can be safely suspended from a platform without impacting its integrity. That type of problem would occur most commonly in conjunction with the hanging culture of mollusks.

Oil and gas platforms are designed to withstand storms, thereby overcoming the anchor dragging and physical loss of support facilities that has occurred in conjunction with some offshore maricultural systems. Because of liability problems, it may be difficult to add a maricultural component to a producing oil or gas platform. In most, and perhaps all instances, it may be necessary to begin the maricultural activity after oil or gas production ceases.

If a platform is converted from production of oil or gas to mariculture, the liability of the owner should be adjusted to reflect the new use. If the production platform is sold by the oil or gas company to another firm that will use the platform for mariculture, the new owner would be expected to carry liability insurance. Since ultimate removal costs would not disappear but would only be deferred during the period a platform was used for mariculture, and since those costs would be a major deterrent to ever establishing a mariculture facility, platform disposal may require innovative advance planning. It may be necessary to establish a mechanism (e.g., an escrow account) by which most or all of the ultimate removal cost would be incurred by the oil and gas company at the time the platform is turned over to the mariculturist. Having the funds in interest-bearing accounts would provide some protection from inflation, ensuring that most or all of the required monies would be available to dismantle the platform when its working life had expired. The mariculturist might share to some extent in funding the escrow account.

Conceptual designs for offshore fish culture facilities have been developed by a number of individuals and groups, and in some cases those concepts have led to the construction and testing of prototype systems. Information on engineering requirements for such facilities is being developed at a rapid pace. Among the primary design considerations for a sea cage or net pen is that the facility should not be so expensive that it cannot be profitable. Some designs that have been proposed, and in some cases constructed and tested, would cost commercial mariculturists millions of dollars (Moffatt, 1991). It is difficult to envision how the fixed facility costs could be amortized.

Offshore mariculture systems must provide a suitable environment for the species under culture, be safe, and be able to endure the rigors of the marine environment (Linfoot et al., 1990). Ease of towing, mooring, and boarding are also considered to be important features. Sveälv (1991b) has recommended round-net bag construction such that individual bags are about 15 m deep and have volumes of 5000 m^3. Standard salmon net pens typically have volumes of 5000 to 15,000 m^3. The maximum practical size may be about 25,000 m^3 (Dahle, 1991a), though most in use today are in the 5000- to 10,000-m^3 range. The use of net pens with large volumes and relatively low stocking densities of 5 to 10 kg/m^3 can result in improved fish growth, better food conversion ratios, lower mortality and incidence of injury, and better disease resistance than when smaller net pens and higher densities are employed. It is likely that the fish sizes and stocking densities presently used for inshore net-pen facilities will form the foundation on which stocking of offshore facilities is based.

Offshore net pens will have to be highly refractory to waves. In addition to causing total system failures, wave energy can cause damage and mortality to the species being cultured (Dahle, 1991a). Moving culture systems into protected waters to avoid storms is an option (National Research Council, 1992), but might often be impractical. It is conceivable that a permitting system could be devised that would allow infrequent and temporary use of protected waters during storm events, but whether sufficient warning could be obtained to allow the safe movement of a facility to protected waters would always be a consideration.

Some manufacturers have taken relatively standard inshore net-pen designs and modified them to provide sufficient resiliency to place the structures in exposed sites. Other companies have come up with structures specifically designed for use in the offshore environment. One example is the OceanSpar™ net-pen system designed by Net Systems of Bainbridge Island, Washington (Loverich, 1991a; Loverich and Croker, 1993; Loverich and Swanson, 1993). An example of that system can be seen in Figure 3.1 to the right of the oil and gas platform. Specialized fittings at mooring points had to be designed because standing fittings quickly failed when exposed to the constant motion of wires and ropes in the open sea (Figure 3.2).

Submerged culture chambers have been discussed in recent years as an alternative method to use when growing fish offshore. The idea has gone beyond the conceptual stage, as a number of styles of submerged culture chambers have been constructed and tested (Muravjev et al., 1993; Okamoto et al., 1993; Dahle et al., 1989b; Dahle and Oltedal, 1990; Willinsky et al., 1991; Willinsky and Champ, 1993; Anonymous, 1995b). Two experimental cage designs, which have been tested

Figure 3.2. Special turnbuckles and other fittings were developed to help the OceanSpar™ net-pen system withstand the random movements of the sea.

Figure 3.3. A large experimental cage designed for attachment to oil and gas platforms in the Gulf of Mexico. Red drum were reared in the cage.

in the Gulf of Mexico, are shown in Figures 3.3 and 3.4. Though not widely in use as yet, submerged net pens or cages would seem to have some significant advantages over floating net pens. Submerged culture chambers would be less impacted by waves, permit avoidance of surface storms without relocating the system, and would be less susceptible to biofouling and corrosion than their floating counterparts (Dahle and Oltedal, 1990; Dahle, 1991a). Submerged net pens or cages would not be subject to damage from floating ice and could be moved up or down in the

Figure 3.4. A second-generation experimental red drum cage to be used in association with oil and gas platforms in the Gulf of Mexico.

water column to avoid noxious plankton blooms or take advantage of appropriate temperature strata when thermal gradients are present (Beveridge, 1987).

The major disadvantages associated with submerged net pens are that the fish farmer cannot readily observe the animals being cultured, and operation of the facility (for example, providing feed) is more difficult than is the case in conjunction with floating net pens (Dahle et al., 1989b). Some species are thought to require periodic access to the surface—perhaps to obtain air to inflate their swim bladders (Moyle and Cech, 1982)—so extended submergence could be a problem (Kuo and Beveridge, 1990). Net pens or cages with submersible capability could be maintained at the surface during calm weather and submerged as required to avoid the problems mentioned.

A seabass and sea bream operation located about 1 km from shore off Corsica employed submersible pens lowered to 15 m during winter and 7 m during the summer (Anonymous, 1993c). The pens were brought to the surface each morning during the summer and submerged in the evening. Because of foul weather the pens were kept submerged throughout the winter. Divers were used to service the pens and feed the fish. The high expenses associated with the heavy use of divers prompted the company to gradually replace the submersible pens with floating ones capable of withstanding waves up to 6 m in height.

Spherical sea cages that are entirely surrounded by netting provide protection from escapement, and, if they are rotated periodically to expose different portions to the atmosphere, biofouling can be controlled. Such cages can be maintained partially submerged during good weather and submerged during storms (Willinsky and Champ, 1993). A 12-m cage has been constructed and tested (Willinsky and Champ, 1993). The design features triangular panels formed into a sphere (basically a geodesic dome design). It has been successfully tested in Canada at offshore locations. Advantages include competitive cost, simple and reliable operation, and the self-cleaning feature for control of biofouling.

A somewhat futuristic concept is a pyramid-shaped net enclosure that would rest on the bottom (Balchen, 1990). The upper portion of the pyramid would contain a bubble of air pumped from the surface to provide the fish with access to atmospheric gases. Feed would automatically flow from a floating surface module. The fish would remain submerged and essentially untended from stocking to harvest.

Some type of surface platform would be required to service submerged net pens (Sveälv, 1991a). The platform could be used to store feed, house automated feed delivery equipment, serve as a work station and perhaps quarters for personnel, and provide a location for automated data acquisition equipment and the necessary power supplies. Cold storage and slaughterhouse facilities could also be included (Dahle et al., 1989a; Sveälv, 1991a). Such a platform should be sufficiently stable to provide a safe working environment for personnel during rough weather (Dahle et al., 1989b). The notion of utilizing oil and gas platforms as support facility structures is compelling.

The FarmOcean International AB cage system is operated as semisubmerged facility (Figure 3.5). Hollow cylinders that make up much of the support structure of each system extend well down into the water column and can be flooded to submerge the net in which the fish are contained while keeping the top of the structure above the water surface. Up to 3 tons of feed can be stored in an automatic feeder mounted atop each unit (Götmalm et al., 1989) so the fish can be fed for at least several days during storm periods when the system cannot be serviced. The feeder and various sensors are operated by an onboard computer. Some of the FarmOcean systems are currently being operated in Scotland, where they have been located near land but in a highly exposed location (Figure 3.5). The structures have reportedly survived waves of at least 6 m (Ronald Roberts, personal communication). FarmOcean has also placed some of its net-pen systems in exposed locations within 2 km from shore in the Mediterranean Sea (McCoy, 1993b).

Submerged experimental sea cages were placed in the Mediterranean Sea off Israel in 1992. By 1995 the cages had survived three winters that included two severe storms (Anonymous, 1995b). The three cages were submerged about 50 m below the surface in water 80 m deep and 11 km offshore. Gilthead sea bream are being reared in the facility (Chamish, 1995). A commercial unit is being designed based on the successes that have been achieved to date.

Net-pen systems are generally constructed in a way that allows them to be towed into position, so in reality they are mobile. In most cases a system will remain anchored in place for periods of one or more years before being relocated. Systems such as one designed by Aqua Systems International have been designed using ship-

Figure 3.5. A semisubmersible FarmOcean International AB cage system used to rear Atlantic salmon in exposed coastal waters in Scotland. The cylindrical object at the top is a feed bin that also supports a computerized meteorological data acquisition and feeding system. The cage is shown in the semisubmerged position. Water is pumped out of the diagonal arms when access is needed during harvesting or stocking.

building technology (Sveälv, 1991a). The basic Aqua Systems design is similar to a ship's hull in shape and is 126.5 m long and 22 m wide. In reality, the design is basically a long raft with the ends configured like the bow and stern of a ship. The design features a flat deck area with large openings in it from which nets can be suspended. The vessel is designed to provide feed storage, living accommodations, and other support functions. Like more standard net-pen systems, it can be anchored in place, but unlike most systems, it can easily be moved as needed. One advantage is that the system is designed to be self-contained and not reliant on shore-based support facilities. The system is designed to house 12 square nets, 14 m on a side, and a triangular net at both the bow and stern. Total volume within the net pens is to be 25,000 m^3. The cost for such a system has been estimated at about $2 million, but at the time of this writing it has not been built.

Exposure of offshore mariculture facilities to routine waves and currents can result in rapid fatigue of netting (Anonymous, 1991a) and other components. In addition, currents can cause deformation of net pens, which may reduce productive capacity or place unacceptable stresses on materials, though some designs can reduce such deformations (Anonymous, 1991a). In addition to fatigue from constant movements associated with currents and waves (Anonymous, 1993b), problems associated with offshore mariculture facilities include rapid electrolysis that can weaken metal components and lead to failures. Mariculture facilities or parts thereof that have been damaged by storms or other mishaps could become hazards to navigation if they drift outside of the delineated culture site. The mariculturist would undoubtedly have to assume liability for such flotsam.

Beveridge (1987) recommended establishing facilities in areas where current speeds are normally less than 50 cm/sec. Not only do currents place stress on net-pen material, tolerance of them varies among culture species. Methods for calculating the forces of currents and waves on mariculture facilities have been published (Aarsnes and Rudy, 1990; Wei et al., 1990). These and various other engineering problems associated with offshore systems can be found in a number of publications (Linfoot and Hall, 1989; Bell and Barr, 1990; Bjerke, 1990; Cairns and Linfoot, 1990; Fearn, 1990; Linfoot et al., 1990; Shaattelid, 1990; Wei et al., 1990; Whittaker et al., 1990; Goudey, 1991; Griffin, 1991; Loverich, 1991; Riley and Mannuzza, 1991; R. J. Taylor, 1991; Loverich and Croker, 1993; Loverich and Swanson, 1993). Dahle (1991a) has considered the problems of providing feed, flotation control, and processing facilities for flotillas of offshore net pens and developed some preliminary designs.

Invertebrates such as scallops, oysters, and mussels could be reared at offshore sites using longline or raft technology with appropriate modifications. The longline technique is widely used in Japan and China, has seen limited use in the United States, and supports an industry in Prince Edward Island, Canada (Judson, 1989). Raft culture has been extensively practiced in Spain for a number of years and was initiated in Portugal in 1976 (Leal, 1989). The technique has also seen limited use in the United States.

Raft and longline culture are now practiced primarily in protected waters, though the techniques could be used offshore in areas where adequate phytoplankton levels occur to provide the natural food required by the mollusks. Surveys of primary productivity would be required as a part of siting activity associated with a mollusk culture operation. While it is conceivable that fish and mollusk polyculture could be developed in areas with low primary productivity—nutrients from the fish production facility would stimulate increased primary production—such operations would be more feasible in bays than in the open ocean.

Unconfined Mariculture

Various approaches for culturing animals without confinement have been put forward over the years. A simple approach is to attract wild fishes to artificial reef structures. Capture is facilitated because the fish are concentrated. Perhaps the other extreme is the notion of using trained porpoises to act as oceanic "cowboys" to maintain a "herd" of fish within a specified area. A more high-tech approach would be to use behavior modification to keep fish within a designated area where they can be captured upon reaching market size.

The use of porpoises for herding fish has never been pursued, but behavior modification is a technique that has received some attention and has, in fact, produced some positive results. Fish have been successfully conditioned to move into a feeding area where they can be concentrated and caught (Øiestad et al., 1986). Music or other sounds broadcast into the water in conjunction with feeding can be used to condition captive fish prior to their release. The sounds might be used at feeding times through the grow-out period; as an alternative the fish could be allowed to forage on their own and then concentrated by sound at the time of harvest. The technique would work best on fish that are sedentary or have restricted home ranges and so remain within the range of the sound that is generated to concentrate them.

Acoustic conditioning has been a subject of interest in Japan since the early 1970s (Conrad Mahnken, personal communication). By the mid-1980s, sea bream were being effectively conditioned to remain in the vicinity of a feeding buoy that emitted a signal 1 or 2 minutes prior to dispensing feed. Fish were fed 5 to 7 times daily at 2-hour intervals. Significant contributions of the conditioned fish to a local fishery were obtained. Conditioning of the sea bream began in a hatchery when the fish were about 30 mm long. After about 80 days of conditioning, the fish were released in the vicinity of the feeding buoy.

Artificial reef technology is well developed, and such reefs have been widely employed as fish-attracting devices in the marine waters of the United States and other nations. In Japan, black rockfish produced in captivity are being stocked over artificial reefs where they eventually recruit into the commercial fishery (Takahashi, 1994).

Nonmechanical barriers such as air bubble curtains, acoustic barriers, and electrodes that discharge pulses of electricity have potential as methods by which fish could be contained within a designated area. Those types of barriers were discussed by Balchen (1991), who suggested such barriers could be "intelligent," meaning they would be activated only when fish were in the vicinity of the boundary. Such approaches are currently theoretical or in the early stages of development.

Petroleum production platforms serve not only as fish-attracting structures (essentially performing as artificial reefs); they also provide substrates for the growth of various species that attach to surfaces. Included are such commercially valuable mollusks as mussels and oysters. Planned harvest of such shellfish is a primitive form of mariculture. One company in California has harvested about 3200 kg of mussels per week from production platforms. That same firm has also been interested in producing oysters and scallops in trays suspended under the platforms (Driessen, 1987). A secondary crop of attached mollusks growing on the platform legs and support beams could become a part of a confinement culture system primarily established for rearing motile species.

CURRENT LEASING POLICIES AND THEIR RELEVANCE TO OFFSHORE MARICULTURE

There is currently no U.S. policy that specifically addresses the development of offshore mariculture. Existing legislation associated with leasing activity in the EEZ is limited to oil and gas leases and leases for mineral exploration and extraction.

Various states with seacoasts have developed procedures with respect to leasing sites for mariculture within coastal waters that fall under state jurisdiction. Those approaches range from banning all but not-for-profit finfish culture (Alaska) to developing comprehensive plans for maricultural development (several states). A summary of current state leasing programs is presented in Table 3.1.

TABLE 3.1. Coastal State Leasing Programs with Respect to Marine Mariculture

State	Leasing Program
Alaska	• Finfish culture is restricted to not-for-profit salmon ranching operations. • A leasing structure has been established for mussels and oysters. • Lease terms are expected to be 10 to 25 years. • Reappraisal and adjustment of rent on leased sites is to be conducted every 5 years.
California	• Leases may be granted to citizens of the state by the Fish and Game Department. • Rent charges are fixed. • Leases are not granted in productive natural shellfish beds or in certain areas that have specifically been excluded. • Leases are granted after a bidding process. • Leases are for contemporary aquaculture production systems. • Twenty-five year leases are available (20 years for kelp farms) and are renewable.
Delaware	• Rent charges are fixed. • Leases can be obtained for traditional shellfish beds only. • Contemporary aquaculture ventures are not accommodated under the current leasing program. • One-year renewable leases are available.
Florida	• Water-column leasing is officially recognized; no more traditional leases are being granted. • Aquacultural leases can be obtained from the Department of Natural Resources to establish contemporary aquaculture ventures. • Rent charges are negotiable. • Exclusive use of the bottom and water column are granted to the extent required by the leasehold. • The maximum initial lease term is 10 years, with successive terms being available on agreement of the parties.
Georgia	• Leases may be provided on state-owned clam or oyster beds by the Department of Natural Resources.
Maine	• Water-column leasing is officially recognized • Rent charges are fixed • Leases may be granted for the water column and bottom areas in subtidal and portions of the intertidal zone by the Commissioner of Maine Resources. • Leases may be as much as 60 ha in size; leases may be granted for as long as 10 years and are renewable with approval from the state.

TABLE 3.1. (continued)

State	Leasing Program
Maryland	• The leaseholder must be a resident of the state. • Rent charges are variable. • Traditional shellfish leases have been granted in the past, but no additional expansion area is available. • Renewable leases for 20 years are available. • Finfish culture has not been an option until recently, when a new policy was developed to allow the issuance of up to 10 permits for experimental net-pen culture on sites up to 0.2 ha in size.
Massachusetts	• Licenses may be granted for the rearing of shellfish below the intertidal zone.
Mississippi	• Oyster leases may be granted by the Marine Conservation Commission.
New Jersey	• The leaseholder must be a resident of the state. • Rent charges are fixed. • Most leases are for traditional clam and oyster production. • Several leases for oyster and clam hatcheries have been granted. • Contemporary aquacultural leasing programs remain to be developed. • One-year renewable leases are available.
New Hampshire	• No lease provisions are currently in place.
New York	• The leaseholder must be a resident of the state. • There is no formal leasing program for contemporary aquaculture ventures, but some special leases have been developed. • Rent charges are variable. • Productive natural shellfish beds may not be the site of aquacultural leases. • Off-bottom culture leases allow the leasehold to purchase and possess submarketable shellfish. • While technically available, no traditional shellfish leases have been granted since the 1930s. • Large areas are in traditional production on private submerged lands. • Leases run for 10 years.
North Carolina	• The Marine Fisheries Commission may lease bottom areas for oyster and clam culture in areas where natural oyster and clam beds do not occur.
Oregon	• Oyster plots can be leased.

(*continued*)

TABLE 3.1. (continued)

State	Leasing Program
Rhode Island	• Water-column leasing is officially recognized. • Rent charges are variable. • Aquacultural leases may be issued by the Coastal Resources Management Council. • Leases provide for exclusive use of the water column above the leased area to the extent required by the leaseholder. • Initial lease term is for 10 years maximum with successive 5-year terms available. • No traditional oyster leases are held at present. • The leases in effect are for mussels and clams.
South Carolina	• The Wildlife and Marine Resources Commission may lease state bottom areas to residents for shellfish culture.
Texas	• The Texas Department of Agriculture has assumed regulatory authority from the Parks and Wildlife Department in 1969.[a] • Aquaculture is not permitted in public waters with the exception of oysters; no oyster culture is presently occurring.[b] • Offshore demonstrations of aquaculture operations have been conducted in federal and state waters.[b]
Virginia	• The leaseholder must be a resident of the state. • The Marine Resources Commission is authorized to lease bottom areas for shellfish culture. • Rent charges are variable. • Leases have been for traditional oyster culture, and no program for contemporary aquaculture has been developed. • Ten-year renewable leases are available.
Washington	• Water-column leasing is officially recognized. • Rent charges are variable. • Both fish and shellfish leases can be obtained. • Both traditional and contemporary aquacultural leases have been granted. • Leases are for 5 years and are renewable.

[a]Texas aquaculture strategic plan and report to the Texas legislature prepared by the Aquaculture Executive Committee, January 1993.

[b]Granvil Treece, Aquaculture Specialist, Texas A&M Sea Grant Program, personal communication, January 1994.

Source: Information from De Voe and Mount (1989), Clay et al. (1981), Anonymous (1993d), DeVoe (1991), Fridley (1993), and Rubino and Wilson (1993).

TABLE 3.2. Characteristics of Aquacultural Leases in 10 Coastal States as of 1989[a]

State	Size (ha)	Duration (yr)	Renewable?	Area under Lease[a]	Number of Leases
California	Variable	25[b]	Yes	880	Not reported
Delaware	20–40	1	Yes	0	Not reported
Florida	Variable	10	Yes	4	Not reported
Maine	≤60	10	Yes	400	66
Maryland	0.4–200	20	Yes	0	Not reported
New York	≥20	10	?	6	—
New Jersey	≤1	1	Yes	Unknown	Not reported
	2–80	1	Yes	Unknown	Not reported
Rhode Island	Variable	Variable	—	67	7
Virginia	0.4–2000	10	Yes	0	Not reported
Washington	0.4–16+	5	Yes	170	40[c]

[a]Excludes traditional shellfish grounds.

[b]20 years for kelp farms.

[c]There were at one time 20 salmon net-pen leases and 20 leases for mussel raft culture, along with a nori culture operation (De Voe and Mount, 1989). At present only about a dozen salmon farms are in operation in Puget Sound, Washington.

Source: Adapted from reference De Voe and Mount (1989).

DeVoe and Mount (1989) reviewed leasing programs in the various coastal states and concluded that "few programs actually meet the needs of the culture industry." Those authors examined the programs in 10 states in some detail. Included in that evaluation were descriptions of the types of leases available, sizes of leased plots, duration of leases, residency requirements, exclusivity of leases, and the amount of land currently being leased in each state. Information on some of those subject areas is summarized in Table 3.2.

Consideration of maricultural leases has perhaps been most extensive in Hawaii, which might provide an example for how other states and the federal government might proceed in the future. Maricultural leases in Hawaii presently come under the state statute that deals with ocean and submerged lands leasing (Title 12, Chapter 190D, Revised Statutes of the State of Hawaii 1980). Leasing is under the auspices of the Board of Land and Natural Resources. Leases will be granted only when the activity is in the public interest; complies with all applicable statutes, ordinances, and rules; and does not interfere with existing activities. Lease arrangements may be negotiated with the board, or a public auction may be held during which the lease would be granted to the highest bidder. The term of each lease and the annual rent are determined by the board. Maricultural leases must specify that the animals or plants being cultured are the property of the leasehold, except that escapees become common property if they cannot be clearly identified as being the property of a leasehold.

Other provisions of current Hawaii leasing policy include penalties of up to $10,000 for mariculturists who operate without having obtained approval of the board. Each day of violation may constitute a separate offense. The policy also outlines civil and criminal liability for damaging, disturbing, or interfering with a maricultural activity.

Aquacultural statutes in Florida, Hawaii, Maine, Maryland, North Carolina, Rhode Island, and Washington have been summarized by Eichenberg and Vestal (1992). Those authors have also looked at submerged-land leasing programs in Delaware, Oregon, and Texas. Among their recommendations to the states are those that could easily be used in the development of an offshore mariculture policy. Some of them are listed below. Italicized words could be replaced with those in brackets to allow the statements to be modified for a federal program in the EEZ.

- promote productive use of *state-owned* [federally controlled] waters as a common resource belonging to and managed for all the people
- develop submerged leasing laws that are specific to *aquaculture* [offshore mariculture]
- include within leasing programs criteria to establish priorities among *aquaculturists* [offshore mariculturists] who are competing for the same site
- establish priorities for competing users for the same site
- provide the *aquaculturist* [offshore mariculturist] with both the right to occupy the site and exclusive rights to the species being cultured

DeVoe and Mount (85) identified five criteria that must be included in a successful state leasing program. The five could also be applied to a federal leasing program.

1. Both the bottom and the water column should be included in the lease.
2. Duration of the lease should be sufficient for startup and establishment of a viable culture operation. It should also provide the *state* [federal government] with flexibility in lease reassignment and termination.
3. The leased area should be adequately protected by law against degraded water quality, theft of culture organisms, and trespassing.
4. Lease fees, bond requirements, and royalties that will be paid by the culturist should be established at the time the lease is granted and not be imposed after the facility is in operation.
5. In order to promote free enterprise and attract investment capital, *states* [the U.S. government] should not require that leaseholds be residents.

Rubino and Wilson (1993) have focused on a variety of regulation issues associated with aquaculture in coastal and inland waters and summarized information on state and federal regulations and leasing policies. Issue areas covered are land use, use of the water column and bottom, water and waste discharge, protection of wild species, culture of nonindigenous species, aquatic animal health, and drug and chemical usage.

Recommendations arising from a workshop held at Woods Hole, Massachusetts, on maricultural policy (Anonymous, 1993a) include one that stresses the need for streamlining the legal process. The responsibilities of the federal government and those of the various states need to be "reconciled and clarified" according to that report.

The laws with respect to aquaculture vary greatly from one nation to another, though most countries claim the bottom area of their territorial waters (Beveridge, 1987). Thus, permits are typically required for floating mariculture facilities because the structures associated with those facilities are moored to the bottom. Little thought seems to have been given to mariculture outside of traditional territorial waters except by Norway and Great Britain, so U.S. policy for mariculture in the EEZ will essentially be a pioneering effort.

In England the seabed under the nation's territorial waters is managed by the Crown Estate. Leases for mariculture provide the right to occupy the seabed in the leased area, and leases are normally 15 years in duration (Side, 1990). A permit from the Department of Transport is also required for fish farms, and there are additional statutes that apply to a few small islands. Acts such as those under which current maricultural leases have been obtained may not apply to waters on the continental shelf outside of the territorial sea. For example, in Scotland there seems to be no policy on offshore mariculture in the EEZ (Niall Bromage, personal communication). The same was true of Norway until 1988, when a commercial firm began a maricultural venture outside of the territorial limit. The government of Norway immediately extended its licensing requirement to cover the entire EEZ (Kjell Naas, personal communication).

Licensing of fish farms in Norway began in 1973 when the Fish Farming Act was passed (Folsom et al., 1992). Under this act, most potential fish farmers were granted licenses to begin operation. In 1981 the government of Norway made it illegal to establish new fish farms or expand existing facilities without permission. The size of salmon farms was later limited to 8000 m^3, which was considered to be small in size. The law was passed to encourage local residents to become involved in fish farming. Further restrictions on the development of large farming operations were instituted in 1985 as the government continued to promote small enterprises in remote areas within territorial waters. Those restrictions were extended to the entire EEZ in 1988. The 1985 act prohibits fishing within 100 m of a mariculture facility or approaching within 20 m. The importation of exotic species is also outlawed by the Fish Farming Act.

If the U.S. government through the Federal Agency Joint Subcommittee on Aquaculture (a body established under the Executive Branch during the Carter administration) is to, "increase the overall effectiveness and productivity of federal aquaculture programs," as indicated in the National Aquaculture Plan (Joint Subcommittee on Aquaculture, 1983), policies facilitating segments of the mariculture industry that will operate in the offshore waters of the nation will be needed.

Options for permitting the establishment of offshore mariculture facilities might include allowing unrestricted development or instituting a licensing or leasing policy. Unrestricted development of offshore mariculture is not an appropriate course of action because of the probable interference of aquatic farms with navigation, commercial fishing, and other activities. Also, the government would relinquish control on the ultimate size of a given facility, which could have environmental consequences if the biomass of maricultured organisms became excessive within a given location. Establishment a licensing or permitting process would provide the mechanism by which orderly development of the offshore aquaculture industry could occur.

As discussed by Wilder (1993a, 1993b, 1993c), policies in the EEZ have commonly been greatly disparate from policy objectives in state waters. Coastal states generally have not shared in the revenues generated from offshore oil and gas leases, and the states have typically created stricter environmental controls on activities within their territorial waters than has been true in the EEZ. Closer cooperation between the coastal states and federal government in the management of EEZ oil and gas exploration and production is needed. That concept can also be extended to other uses of the EEZ, including mariculture. Three potential approaches have been put forward for expanding state jurisdiction in the EEZ (Wilder, 1993c):

- increase the percentage of revenue shared by the federal government with the coastal states.
- extend the authority of coastal states to provide them with shared responsibility for offshore oil and gas activity.
- extend exclusive state control to 19.2 km from the existing 4.8 km.

Of these three, Wilder (1993c) prefers a hybrid approach under which joint state-federal authority would occur in the zone from 4.8 to 19.2 km. The purpose would be to avoid the state-federal tension and stalemate now evidenced by drilling moratoria.

Mariculture can be expected to move offshore incrementally, and if development of the industry in conjunction with oil and gas production platforms becomes a reality, operations may exist both within state waters and in the EEZ. Thus, it should be anticipated that coastal states will seek to exercise some authority over any maricultural development in the adjacent EEZ. Significantly, cooperation between coastal states as well as between the states and the federal government will likely be necessary to effect successful aquaculture development since some lease sites may straddle the waters under the jurisdiction of two states.

Because of the high risks involved with aquaculture in open water, the expenses associated with establishing facilities that can survive in exposed marine environments, and the need for additional research to provide the foundation on which higher-value species can be reared, the profitability of offshore aquaculture may be limited. In no case will profits match or even approach those that have accrued from productive oil and gas leases. Moreover, the establishment of high lease or license fees in conjunction with offshore aquacultural development could mean the difference between profit and loss.

Oil and gas leasing procedures include entertaining bids for exploration in identified areas, so that the highest bidder is awarded the lease for a period of years and with renewal options. While it is conceivable for the government to establish a parallel program with respect to offshore aquaculture, costs of surveys conducted by the government to identify appropriate sites and the expenses associated with establishing a bidding process would probably not be recovered, as bids would be extremely low relative to those that have been associated with oil and gas exploration.

While current oil and gas leasing policy is not an entirely suitable model for development of an offshore aquacultural policy, there are elements of that policy that could be utilized. A fee structure for aquacultural leases based on royalties would provide relief to the aquaculturist during the startup period but would eventually mean income to the government should the venture become profitable. A bidding process would be appropriate in instances where two or more applicants become interested in the same site. Current oil and gas policy requires cooperation by adjacent states, which would be of benefit if the states worked toward an appropriate development policy, but which could be detrimental if the requirement were used by a state to entirely block maricultural development in the offshore area.

Whether offshore aquacultural policy involves leasing or licensing, the process should be as simple as possible. The process should also be clear so that the prospective mariculturist knows exactly what steps to follow in order to obtain the necessary permits and the lease or license. The concept of one-step shopping deserves strong consideration.

Licensing programs could be undertaken by the states or by the federal government. Since some offshore facilities may overlap the two jurisdictions, it might be most reasonable to allow the states to administer licenses under regulations developed jointly. Federal government involvement might be appropriately housed in the National Marine Fisheries Service or in the U.S. Department of Agriculture. Coordination could be provided through the Joint Subcommittee on Aquaculture. Most coastal states now have aquacultural coordinators, located in many instances in the state's department of agriculture. Coordination of leasing or licensing of offshore mariculture operations through those state aquacultural coordinators would seem to be a logical extension of current activities.

Marine sanctuaries that would otherwise restrict aquaculturists could be licensed for the production of fish or other organisms to be used for enhancement or to assist in the recovery of threatened or endangered species. Once a recovery program was completed, the license could be terminated. Resources for such activities could come from federal programs and user fees.

Coastal states have tended to place environmental protection above other considerations when promulgating policies associated with development in territorial waters (Wilder, 1993c). The National Research Council (1990) has found that the five most often mentioned informational needs of the states with respect to the EEZ are management of biological resources, mineral resources management, environmental assessment, shoreline management, and oil and gas development. Of less importance to the states are data on pipelines, cables, ocean energy development, geohazards, cultural and recreational interests, and military uses. Aquaculture is not a priority in the states. That could change if an equitable policy for leasing or licensing could be developed. There is certainly the potential for economic gain within coastal states that promote marine aquacultural development. Jobs are created and taxes are paid. Support industries can expand, leading to additional revenue to the states.

Policies developed for the permitting of offshore aquaculture should include appropriate penalties for vandalism and theft. Those penalties should be commensurate with similar activities conducted on state and federally owned or controlled lands.

Operators of offshore aquacultural facilities should be liable for negligence such as failure to replace navigational aids that are not operating properly. Safe working conditions should be provided for personnel associated with offshore facilities, though it must be recognized that the environment in which such farms are established is by its very nature hazardous. The permitting or licensing process could stipulate the types of bonds and liability insurance required for each facility.

The U.S. Merchant Marine Act as amended in October 1992 may provide some mortgage and insurance relief for mariculturists (McCoy, 1993b). Under the amended act, fisheries facilities eligible for federal loan and mortgage insurance guarantees would include mariculture. Insurance for catastrophic losses of facilities and crops would not seem to be covered but would have to be obtained from private insurance companies. The premiums for insurance will undoubtedly be higher for offshore mariculturists than for their inshore counterparts. Premium rates may be so high that profit margins are eliminated, so it may be in the best interest of the offshore mariculturists to accept the risks and face periodic crop failures that are not protected by insurance.

The U.S. Congress can be expected to develop an offshore aquacultural policy in response to interest in the private sector to establish facilities in the EEZ. What that policy will look like remains to be seen, but clearly there are a number of issues that must be faced, and the process will likely require a good deal of time and debate.

CONCLUDING REMARKS

The future of marine aquaculture has been envisioned as moving in two directions: into recirculating water systems and offshore (National Research Council, 1992). While proponents of recirculating water systems, particularly those designed for freshwater species, have been around for more than two decades, the technology for constructing economically viable systems has remained elusive. Recent advances in the technology associated with biofiltration, solids handling, ozonation, and other system components and functions would seem to bode well for the establishment of increasing numbers of profitable systems.

Offshore mariculture now appears to be at the stage recirculating systems were a decade ago. Interest is keen, but many engineering and logistical problems remain to be solved. As more and more scientists and engineers become interested in the development of offshore mariculture systems, resolution of what have seemed to be insurmountable problems can be anticipated. Costs will always be high in conjunction with establishing and operating offshore mariculture facilities, but with increasingly stringent environmental protection regulations, in conjunction with shore-based and nearshore sites, there may be no realistic alternative to moving offshore.

It will be interesting to watch as offshore aquaculture develops over the next decade or two. With diminishing supplies of wild fish and shellfish and an increasing human population, there will be many new opportunities for aquaculture to make a contribution toward filling an increasing gap between supply and demand. Offshore mariculture can play a role, and perhaps in the future a major role, in that effort.

REFERENCES

Aarsnes, J. V., and H. Rudy. 1990. Current forces on cage, net deflection. pp. 137–152, In: Engineering for offshore fish farming. Thomas Telford, London.

Ackefors, H. 1992. Pollution loads from land-based and water-based aquaculture systems. Paper presented in the Workshop on Fish Farm Effluents—and Their Control in EC Countries, November 23–25, Hamburg, Germany. 37 p.

Ackefors, H. 1993. Waste management. Paper presented at the meeting on Aquaculture and Marine Environment. The shaping of public policy, August 30–September 1, Woods Hole, Mass. 8 pp.

Anonymous. 1991a. Fatigue life is important. Fish Farm. Int., 18(12): 28.

Anonymous. 1991b. Nets cage passes tough test in Gulf of Mexico. Fish Farm. Int., 18(12): 28–29.

Anonymous. 1993a. Aquaculture and the marine environment: the shaping of public policy. Summary of a workshop held at the marine Biological Laboratory, August 30–September 1, Woods Hole, Mass. 30 p.

Anonymous. 1993b. Netting and net pen design for offshore cages. World Aquacult., 24(1): 62–65.

Anonymous. 1993c. Sea farm off Corsica changes its cages. Fish Farm. Int., 20(10): 28–29.

Anonymous. 1993d. Net pen culture. Aquacult. Mag., September/October, p. 102.

Anonymous. 1995a. Aquaculture in France: farming the sea. Aquacult. Mag., 21(1): 39–42.

Anonymous. 1995b. Sink *and* swim. Watermark, 1: 4.

Balchen, J. G. 1990. The state of the art in offshore fish farming. pp. 211–219, In: Engineering for offshore fish farming. Thomas Telford, London.

Balchen, J. G. 1991. Possible roles of remotely-operated underwater vehicles (ROV) robotics in future mariculture. pp. 373–385, In: G. N. Hirata, K. R. McKinley, and A. W. Fast (Eds.). Workshop on Engineering Research Needs for Off-shore Mariculture Systems, September 26–28, Honolulu. Hawaii Natural Energy Institute, Honolulu.

Barker, C. J. 1990. Classification society rules for fish farms. pp. 31–38, In: Engineering for offshore fish farming. Thomas Telford, London.

Bell, A. K., and A. G. Barr. 1990. Computerised mathematical modelling in the assessment of fish-farming sites. pp. 49–61, In: Engineering for offshore fish farming. Thomas Telford, London.

Beveridge, M. C. M. 1987. Cage aquaculture. Fishing News Books, Farnham, England. 352 p.

Beveridge, M. C. M., and J. F. Muir. 1982. An evaluation of proposed cage culture in Loch Lomond, an important reservoir in central Scotland. Can. Water Res. J., 7: 107–113.

Bingham, A. 1991. The deeper the better (cage fish culture, Hong Kong). Far East Agricult., 1991: 27–28.

Bjerke, K. S. 1990. Offshore fish-farming platforms—development, design, construction and operation: the SEACON and SULAN concepts. pp. 153–169, In: Engineering for offshore fish farming. Thomas Telford, London.

Cairns, J., and B. T. Linfoot. 1990. Some considerations in the structural engineering of sea-cages for aquaculture. pp. 63–77, In: Engineering for offshore fish farming. Thomas Telford, London.

Chamish, B. 1995. Former fishing vessel runs mid-sea farm. Fish Farm. Int., 22(8): 27.

Clay, G. S., S. Broder, R. Turner, D. S. Kitoaka, G. L. Rhodes, and D. K. Yamase. 1981. Ocean leasing for Hawaii. Hawaii Department of Planning and Economic Development, Honolulu. 124 p.

Cooper, C. K., G. Z. Forristall, R. C. Hamilton, and C. C. Ebbesmeyer. 1993. Utilization of offshore oil platforms for meteorological and oceanographic measurements. Mar. Tech. Soc. J., 27: 10–23.

Dacey, J. W. H., and D. J. Cooper. 1993. Dynamics for dimethylsulfide in coastal waters and the marine atmosphere: a need for platform observations. Mar. Tech. Soc. J., 27: 72–77.

Dahle, L. A. 1991a. Exposed fish farming: biological and technical design criteria and possibilities. pp. 23–39, In: G. N. Hirata, K. R. McKinley, and A. W. Fast (Eds.). Workshop on Engineering Research Needs for Off-shore Mariculture Systems, September 26–28, Honolulu. Hawaii Natural Energy Institute, Honolulu.

Dahle, L. A. 1991b. Offshore aquaculture technology—possibilities and limitations. pp. 83–84, In: N. DePauw and J. Joyce (Comp.). Aquaculture and the environment. European Aquaculture Society Spec. Publ. 14. Bredene, Belgium.

Dahle, L. A., and G. Oltedal. 1990. Norwegian research and industrial development of floating structures for salmon fish farming. pp. 185–196, In: Engineering for offshore fish farming. Thomas Telford, London.

Dahle, L. A., L. Jørgensen, K. Vangen, and J. V. Aarsnes. 1989a. Basic requirements of an off-shore control and feeding station for fish-cage flotillas: a theoretical evaluation. pp. 1009–1017, In: M. De Pauw, E. Jaspers, H. Ackefors, and M. Wilkins (Eds.). Aquaculture—a biotechnology in progress. European Aquaculture Society, Bredene, Belgium.

Dahle, L. A., L. Jørgensen, K. Vangen, and J. V. Aarsnes. 1989b. An underwater offshore sea cage. pp. 1001–1007, In: M. De Pauw, E. Jaspers, H. Ackefors, and M. Wilkins (Eds.). Aquaculture—a biotechnology in progress. European Aquaculture Society, Bredene, Belgium.

DeVoe, M. R. 1991. Regulatory aspects of aquaculture development. pp. 135–164, In: J. A. Hargreaves and D. E. Alston (Eds.). Status and potential of aquaculture in the Caribbean. World Aquaculture Society, Baton Rouge, La.

DeVoe, M. R., and A. S. Mount. 1989. An analysis of ten state aquaculture leasing systems: issues and strategies. J. Shellfish Res., 8: 233–239.

Dokken, Q. 1993. Flower gardens ocean research project: using offshore platforms as research stations. Mar. Tech. Soc. J., 27: 45–50.

Donaldson, E. M. 1991. Application of biotechnology to biological problems associated with off-shore mariculture. pp. 87–99, In: G. N. Hirata, K. R. McKinley, and A. W. Fast (Eds.). Workshop on Engineering Research Needs for Off-shore Mariculture Systems, September 26–28, Honolulu. Hawaii Natural Energy Institute, Honolulu.

Driessen, P. K. 1987. Oil rigs: biology, mariculture, drilling muds, rigs-to-reefs. pp. 3605–3620, In: O. T. Magoon, H. Converse, D. Miner, L. T. Tobin, D. Clark, and G. Domurat (Eds.). Coastal Zone '87. Proceedings of the Fifth Symposium on Coastal and Ocean Management, May 26–29, Seattle. American Society of Civil Engineers, New York.

Edwards, D. J. 1978. Salmon and trout farming in Norway. Fishing News Books, Farnham, England. 195 p.

Eichenberg, T., and B. Vestal. 1992. Improving the legal framework for marine aquaculture: the role of water quality laws and the public trust doctrine. Terr. Sea J., 2: 339–404.

Fearn, P. T. 1990. The development of an offshore fish cage. pp. 107–117, In: Engineering for offshore fish farming. Thomas Telford, London.

Findlay, G. 1990. Insurance and risk control. pp. 1–3, In: Engineering for offshore fish farming. Thomas Telford, London.

Folsom, W., D. Altman, A. Manuar, F. Nielsen, T. Revord, E. Sanborn, and M. Wildman. 1992. Norway. pp. 93–132, In: W. B. Folsom (Ed.). World Salmon Culture. National Marine Fisheries Service, Silver Spring, Md.

Fridley, R. B. 1991. The opportunities for engineering and technology in addressing the environmental, institutional, and economic constraints of marine aquaculture in the U.S. pp. 41–54, In: G. N. Hirata, K. R. McKinley, and A. W. Fast (Eds.). Workshop on Engineering Research Needs for Off-shore Mariculture Systems, September 26–28, Honolulu. Hawaii Natural Energy Institute, Honolulu.

Fridley, R. B. 1993. Marine aquaculture—opportunities for growth. Sea Technol., August, pp. 10–20.

Götmalm, Ö. A., T. Boberg, and T. L. Sveälv. 1989. The development of a semi-submersible offshore fish farm. p. 1019, In: M. De Pauw, E. Jaspers, H. Ackefors, and M. Wilkins (Eds.). Aquaculture—a biotechnology in progress. European Aquaculture Society, Bredene, Belgium.

Goudey, C. A. 1991. Design and construction of retention and attachment devices for offshore mariculture. pp. 121–130, In: G. N. Hirata, K. R. McKinley, and A. W. Fast (Eds.). Workshop on Engineering Research Needs for Off-shore Mariculture Systems, September 26–28, Honolulu. Hawaii Natural Energy Institute, Honolulu.

Gowen, R. J., and A. Edwards. 1990. The interaction between physical and biological processes in coastal and offshore fish farming: an overview. pp. 39–47, In: Engineering for offshore fish farming. Thomas Telford, London.

Griffin, R. B. 1991. Corrosion of materials used in off-shore construction. pp. 131–156, In: G. N. Hirata, K. R. McKinley, and A. W. Fast (Eds.). Workshop on Engineering Research Needs for Off-shore Mariculture Systems, September 26–28, Honolulu. Hawaii Natural Energy Institute, Honolulu.

Jeffress, G. A., E. G. Masters, and P. R. Michaud. 1993. The Texas coastal ocean observation network and oil platforms in the Gulf of Mexico. Mar. Tech. Soc. J., 27: 51–55.

Joint Subcommittee on Aquaculture. U.S. Congress. 1983. National aquaculture development plan. U.S. Government Printing Office, Washington, D.C. Vol. I: 67 p.

Judson, W. I. 1989. Mussel culture in Prince Edward Island. 1989. pp. 336–339, In: M. De Pauw, E. Jaspers, H. Ackefors, and M. Wilkins (Eds.). Aquaculture—a biotechnology in progress. European Aquaculture Society, Bredene, Belgium.

Kuo, C., and M. C. M. Beveridge. 1990. Mariculture: biological and management problems, and possible engineering solutions. pp. 171–184, In: Engineering for offshore fish farming. Thomas Telford, London.

Lazzari, A., and M. DiBitetto. 1993. Demonstrative offshore farming project in Italy. p. 304, In: M. Carrillo, L. Dahle, J. Morales, P. Sorgeloos, N. Svennevig, and J. Wyban (Eds.). From discovery to commercialization. European Aquaculture Society Spec. Publ. 19. Bredene, Belgium.

Leal, A. M. 1989. Raft culture of mussels, *Mytilus edulis* L., in Portugal. pp. 351–357, In: M. De Pauw, E. Jaspers, H. Ackefors, and M. Wilkins (Eds.). Aquaculture—a biotechnology in progress. European Aquaculture Society, Bredene, Belgium.

Lewis, J. M. 1993. Challenges and advantages of collecting upper-air data over the Gulf of Mexico. Mar. Tech. Soc. J., 27: 56–65.

Linfoot, B., and M. Hall. 1989. Model testing of single-point mooring systems for sea-cage flotillas. pp. 1029–1036, In: M. De Pauw, E. Jaspers, H. Ackefors, and M. Wilkins (Eds.). Aquaculture—a biotechnology in progress. European Aquaculture Society, Bredene, Belgium.

Linfoot, B. T., J. Cairns, and M. G. Poxton. 1990. Hydrodynamic and biological factors in the design of sea-cages for fish culture. pp. 197–210, In: Engineering for offshore fish farming. Thomas Telford, London.

Loverich, G. 1991a. A discussion of two offshore experiments with OceanSpar™ net pens. Bull. Aquacult. Assoc. Can., 91(3): 108–110.

Loverich, G. F. 1991b. Materials for retention and attachment devices for off-shore mariculture systems. pp. 101–120, In: G. N. Hirata, K. R. McKinley, and A. W. Fast (Eds.). Workshop on Engineering Research Needs for Off-shore Mariculture Systems, September 26–28, Honolulu. Hawaii Natural Energy Institute, Honolulu.

Loverich, G., and T. Croker. 1993. Offshore sea farms: 32 months experience with OceanSpar™ systems. p. 411, In: M. Carrillo, L. Dahle, J. Morales, P. Sorgeloos, N. Svennevig, and J. Wyban (Eds.). From discovery to commercialization. European Aquaculture Society Spec. Publ. 19. Bredene, Belgium.

Loverich, G. F., and K. T. Swanson. 1993. Offshore sea farms: 25 months of experience. pp. 241–253, In: J.-K. Wang (Ed.). Techniques for modern aquaculture. American Society of Agricultural Engineers, St. Joseph, Mich.

McCoy, H. D. II. 1993a. Open ocean fish farming: part one. Aquacult. Mag., September/October, pp. 66–74.

McCoy, H. D. II. 1993b. Open ocean fish farming: part two. Aquacult. Mag., November/December, pp. 60–67.

Miget, R. 1995. Assessment of an offshore net pen finfish production system in the Gulf of Mexico. pp. 261–269, In: Proceedings of the PACON Conference on Sustainable Aquaculture '95, June 11–14, Honolulu. PACON International, Hawaii Chapter, Honolulu.

Moffatt, B. 1991. Seacage design technology. Bull. Aquacult. Assoc. Can., 91(1): 16–21.

Moyle, P. B., and J. J. Cech, Jr. 1982. Fishes: an introduction to ichthyology. Prentice-Hall, Englewood Cliffs, N.J. 593 p.

Muir, J. F., and D. J. Baird, 1993. Environmental management of aquaculture development, Cyprus. FAO Tech. Rep. FITCPCYP9152. FAO, Rome. 84 p.

Muravjev, W., L. Bugrov, and L. Bougrova. 1993. Underwater aquaculture technology in offshore areas. p. 417, In: M. Carrillo, L. Dahle, J. Morales, P. Sorgeloos, N. Svennevig, and J. Wyban (Eds.). From discovery to commercialization. European Aquaculture Society Spec. Publ. 19. Bredene, Belgium.

National Research Council. 1985. Disposal of offshore platforms. National Academy Press, Washington, D.C. 76 p.

National Research Council. 1990. Interim report of the committee on exclusive economic zone information needs. Coastal States and Territories. National Academy Press, Washington, D.C. 46 p.

National Research Council. 1992. Marine aquaculture: opportunities for growth. National Academy Press, Washington, D.C. 290 p.

Øiestad, V. T., A. Pedersen, A. Bjordal, and P. G. Kvenseth. 1986. Automatic feeding and harvesting of juvenile Atlantic cod (*Gadus morhua* L.) in a pond. pp. 199–204, In: J. G. Balchen and A. Tysso (Eds.). Automation and Data Processing in Aquaculture. IFAC Symposium, August 18–24, Trondheim, Norway. Moberg and Helli, Oslo.

Okamoto, M., O. Nagahama, G. Kai, and N. Takatsu. 1993. Development of mid-layer floating type marine cultivation facility. pp. 132–136, In: J. S. Chung, B. J. Natvig, B. M. Das, and L.-Y. Cheng (Eds.). Proceedings, Third International Offshore and Polar Engineering Conference, Singapore. International Society of Offshore and Polar Engineers, Golden, Colo.

Parametrix, Inc. 1990. Fish culture in floating net-pens. Final programmatic environmental impact statement prepared for the Washington Department of Fisheries, Olympia. 161 p.

Petrell, R. J., and K. Mazhari. 1992. Technical and economical feasibility of kelp culture near a salmon netpen farm. Aquaculture '92: growing toward the 21st century. Abstracts of the Aquaculture '92 meeting, May 21–25, Orlando, Fla. p. 185.

Plante, J. M., and S. Jones. 1994. Atlantic Ocean salmon farm seeks go-ahead. Fish Farm. News, 2(4): 8.

Pullin, R. S. V., H. Rosenthal, and J. L. Maclean. 1993. Environment and aquaculture in developing countries. International Center for Living Aquatic Resources Management, Manila. 359 p.

Revised Statutes of the State of Hawaii. 1980. Title 12, Honolulu. 190 p.

Riley, J. G., and M. Mannuzza. 1991. Hydrodynamic forces and the design of fish cage mooring systems. pp. 213–237, In: G. N. Hirata, K. R. McKinley, and A. W. Fast (Eds.). Workshop on Engineering Research Needs for Off-shore Mariculture Systems, September 26–28, Honolulu. Hawaii Natural Energy Institute, Honolulu.

Roscigno, P. F., and M. C. Kennicutt II. 1993. Marine ecosystems: assessing the impacts of chronic contamination from offshore platforms. Mar. Tech. Soc. J., 27: 66–71.

Rubino, M. C., and C. A. Wilson. 1993. Issues in aquaculture regulation. Bluewaters, Inc., Bethesda, Md. 72 p.

Secretan, P. A. D. 1980. Insurance and risk management as related to cage culture. pp. 126–136, In: Proc. IMF Cage Fish Rearing Symposium, Reading University, March 26–27. Janssen Services, London.

Shaattelid, O. H. 1990. Model tests with flexible, circular floats for fish farming. pp. 93–106, In: Engineering for offshore fish farming. Thomas Telford, London.

Side, J. 1990. Controls and legal provisions governing offshore fish-farm developments. pp. 11–29, In: Engineering for offshore fish farming. Thomas Telford, London.

Smith, B. L. 1990. Risk assessment of offshore fish-farming cages. pp. 5–10, In: Engineering for offshore fish farming. Thomas Telford, London.

Sveälv, T. 1988. Inshore versus offshore farming. Aquacult. Eng., 7: 279–287.

Sveälv, T. L. 1989. Inshore versus offshore farming. pp. 253–258, In: M. De Pauw, E. Jaspers, H. Ackefors, and M. Wilkins (Eds.). Aquaculture—a biotechnology in progress. European Aquaculture Society, Bredene, Belgium.

Sveälv, T. 1991a. Moving off-shore fish farms into the open sea: strategies and technologies. pp. 239–263, In: G. N. Hirata, K. R. McKinley, and A. W. Fast (Eds.). Workshop on Engineering Research Needs for Off-shore Mariculture Systems, September 26–28, Honolulu. Hawaii Natural Energy Institute, Honolulu.

Sveälv, T. 1991b. Strategies and technologies in offshore farming. Fish. Res., 10: 329–349.

Takahashi, K. 1994. Sea ranching the black rockfish, *Sebastes schlegeli,* in Japan. pp. 35–36, In: T. Nosho (Ed.). Proceedings of the Marine Culture and Enhancement Workshop, October 4–6, 1993, Seattle, Wash. University of Washington Sea Grant Program, Seattle.

Taylor, C. D., B. L. Howes, and K. W. Doherty. 1993. Automated instrumentation for time-series measurement of primary production and nutrient status in production platform-accessible environments. Mar. Tech. Soc. J., 27: 32–44.

Taylor, R. J. 1991. Anchoring, mooring, and positioning. pp. 157–212, In: G. N. Hirata, K. R. McKinley, and A. W. Fast (Eds.). Workshop on Engineering Research Needs for Off-shore Mariculture Systems, September 26–28, Honolulu. Hawaii Natural Energy Institute, Honolulu.

Wei, G., V. G. Panchang, and B. R. Pearce. 1990. Some numerical models for predicting wave conditions around aquaculture sites. pp. 119–135, In: Engineering for offshore fish farming. Thomas Telford, London.

Whittaker, T. J. T., A. K. Bell, and M. R. Shaw. 1990. Model testing of a cage system for the offshore environment. pp. 79–92, In: Engineering for offshore fish farming. Thomas Telford, London.

Wiebe, P. H., D. D. Moran, R. Knox, C. B. Miller, and J. A. McGowan. 1993. Long-range needs for deep-sea platforms: the deep-sea observatory concept. Mar. Tech. Soc. J., 27: 24–31.

Wilder, R. J. 1992. The three-mile territorial sea: its origins and implications for contemporary offshore federalism. Virginia J. Int. Law, 32: 681–746.

Wilder, R. J. 1993a. Cooperative governance, environmental policy, and management of offshore oil and gas in the United States. Ocean Dev. Int. Law, 24: 41–62.

Wilder, R. J. 1993b. Is this holistic ecology or just muddling through? The theory and practice of marine policy. Coastal Manage., 21: 209–224.

Wilder, R. J. 1993c. Sea-change from Bush to Clinton: setting a new course for offshore oil development and U.S. energy policy. UCLA J. Environ. Law Pol., 11: 131–173.

Willinsky, M. D., and M. A. Champ. 1993. Offshore fish farming: Reversing the 'Oceanic Dustbowl.' Sea Technol., August, pp. 21ff.

Willinsky, M. D., D. R. Robson, W. J. Vangool, R. A. Fournier, and J. H. Allen. 1991. Design of a spherical submersible self-cleaning aquaculture system for exposed sites. pp. 317–336, In: G. N. Hirata, K. R. McKinley, and A. W. Fast (Eds.). Workshop on Engineering Research Needs for Off-shore Mariculture Systems, September 26–28, Honolulu. Hawaii Natural Energy Institute, Honolulu.

4 Aquaculture, Pollution, and Biodiversity

JOHN E. BARDACH

INTRODUCTION

Aquaculture sites can become polluted and cause pollution, in turn, mostly of themselves. Broadly speaking, pollution is any human-caused change in the physical, chemical, or biological properties of the environment, be that air, water, soil, ocean, lake, forest or a city. It usually causes socioeconomic ill effects. As the diversity of plant and animal species in any part of the environment is one of its biological properties that can influence aquaculture or be influenced by it, diversity is treated in this same chapter along with pollution. Aside from its relevance for long-term equilibrium in the biosphere, biodiversity is a goal in the conservation of natural areas. It is also important for plant and animal breeding and a source of pharmaceutical compounds (Eldredge et al., 1996).

Consideration and concern for the environment have greatly increased during the last three decades of the 20th century. This attitude has not only prevailed in technically developed nations but, according to a wide public opinion poll (Bloom, 1995), is beginning to be felt also in parts of the world where income levels might not lead us to expect it. There appears to be apprehension not only about the quality of air or water but also about vanishing and endangered species, or, differently put, there is concern about biodiversity.

It is clear and readily demonstrable that changes in land use—primarily in agriculture, extractive forestry, and urbanization—have led to species extinction and an environment of relative uniformity. In fact, recent extinction rates in certain groups appear to be 100 to 1000 times prehuman levels, and selective acceleration is also likely in the next century (Arrow et al., 1995).

The history of the Hawaiian Islands, with its vanishing native bird fauna and a reduction in the variety in stream fishes and invertebrates (Eldrege and Miller, 1995), is a case in point. While the marine realm everywhere has until now been less affected by various human influences than the environment on land, there is ample suspicion that the husbandry and selective rearing of certain species—aquaculture

Sustainable Aquaculture, Edited by John E. Bardach
ISBN 0-471-14829-6

in short—has begun to affect marine biodiversity also (Eldredge et al., 1996). It should be noted that there are a great many destructive influences on plant and animal life in water, many of them originating on land, with pollution probably being the most important (Laws, 1994). So far, aquaculture has had only minor effects, but as it develops further in variety and extent, it will have disproportionately greater effects both on the environment and its own culture operations. An examination of these interactions is therefore germane to the sustainability of both aquaculture and the environment.

The last few decades have seen many efforts to curb aquatic pollution; it is public knowledge that these have not always been successful, even in developed nations. In parts of the developing world, especially where rapid industrialization is in progress, effluent management policies, mostly the enforcement of laws and regulations, such as exist, are often not followed, mostly because of lack of manpower. But there abound examples of the cleanup of coastal waters in which aquaculture may or may not be or have been practiced. Minamata Bay in Japan, famous for its mercury poisoning, and Kaneohe Bay in Hawaii with its extreme enrichment before sewage diversion, are just two of them (Laws, 1994). Another obstacle to dealing with water pollution that will still remain for a while is insufficient knowledge of the extremely complex ecological relationships among microscopic and macroscopic members of aquatic food webs.

In coastal waters and in lakes, aquaculture installations are more susceptible to pollution damage than husbandry in ponds. Open waters receive many additives from single or multiple sources, and the farmer has no recourse in law against changes to "his" water caused by many additives and agents, which are often far away. A rare exception here is British trout streams, which are, *de facto*, owned; damage to them is therefore more easily brought to court. Prevention and/or control of pollution of cages, net enclosures, and suspension or attachment sites has to rely on public control measures (laws and ordinances) and their enforcement. That these vary, most generally with level of economic development, is almost self-evident, as is the fact that pollution is felt not only by aquaculture but by other uses of the affected waters. Even though clean water is one of the prerequisites of aquaculture, polluted conditions often arise gradually; these are noticed as the uses of the coastal zone increase, usually as new industries are established, perhaps upstream in a river. This was the case for a certain time with the Rhine River in Europe, before international compacts took effect (Laws, 1994).

Other examples of population-density-dependent nutrient enrichment and pollution conditions are certain bays around Hong Kong and the Baltic, where shallow water and poor flushing have aggravated damages to fish in cages (Gowen and Rosenthal, 1993; Rosenthal, 1985).

Most additives to water that cause pollution are the nutrient chemicals: nitrogen and phosphorus. They enhance algal growth in the waters surrounding the cultivars, eventually depleting oxygen at night and giving rise to algal blooms; some of these cause the so-called red tides, which are the sources of certain poisons with potentially grave public health liabilities.

The poisons enter mollusks, shrimps, and plankton-feeding fish through the food chain. Maclean (1993) has listed seven different kinds to be found under severe enrichment conditions of coastal waters in developing countries. Carnivorous fish reared in cages or net enclosures and fed with pellets or trash fish caught at large are less susceptible to these poisons, but they can succumb nevertheless. The poisons are postulated to arise from the bacteria associated with the algae; although these poisons do not affect the bacteria's temporary shellfish and finfish hosts, they can make people very ill, even leading to death, as is the case with paralytic shellfish poison (PSP). These public health effects of pollution are among several that point to the importance of control over postharvest and marketing conditions of the products of aquaculture. Concerning the possible harm caused by algal blooms, we should also note that not all blooms are due to coastal enrichment; a type of bloom with different dominant species may be connected with a periodically recurring ENSO (El Niño Southern Oscillation) conditions (Maclean, 1993).

Since red tides are the most serious effect of pollution by nutrient enrichment, various techniques have been proposed to deal with them. Thorough knowledge of hydrologic conditions at the culture site certainly seems to be a condition for applying these techniques. Water that is stagnant or has slow tidal exchange should simply be avoided. Placing sleeves around the cages to keep the algae out make it necessary to aerate the water or bring water up from below, since the algae are densest on the surface of the water. By the same token, the cages may be lowered at certain times if they are closed on top.

Past experiences with industrial effluents and the specific toxicities of certain metals and organic chemicals suggest that aquaculture should not be sited in the vicinity of industries. The trend worldwide will eventually be to eliminate the chemicals in such effluents. The caveat here is that it may take years to notice their effects in plants and animals and hence on people exposed to them. The mercury poisonings in Japan (e.g., Minamata Bay, already mentioned) and in other locations are cases in point (Laws, 1994). Industrial catastrophies and oil spills are rapid-onset disasters; how to prevent them or deal with them after the fact is the subject of much writing (National Research Council, 1989) in which aquaculture figures only occasionally and only as one among many affected entities.

The population densities used in aquaculture are a major factor in producing pollution. Ponds, cages, pens, and rafts with attachment devices (ropes and so on) for mollusks all have thousands of individuals, sometimes of an introduced species, in the volume or space where in the wild we would normally find perhaps a dozen individual animals. For most intensive cultures this may amount to per-hectare productions of 50 tons of biomass or more. The effluent water, or the water surrounding the installations if there is only tidal exchange, is thus often greatly enriched by the metabolism of the cultivars or by the decomposition of unused food (Table 4.1).

Intensive shrimp culture, for instance, adds to the water between 63% and 78% of the nitrogen contained in the feed of shrimp and between 76% and 86% of the phosphorus. The amount of loss of this material into the water depends on the food

TABLE 4.1. Effects of Fish-Farm Effluents on Receiving Waters in Finland

Effect	Number of Cases
Eutrophication	22
Increased phosphorus content	15
Increased bacterial counts	11
Lowered oxygen content	9
Algal blooms	8
Sewage fungus and settle solids	5
Increase in chlorophyll a	4
Increased macrophytes	3
Increased water turbidity	2
Smell	2
Bad fish taste	2
Water not potable	2
Restricted use of potable water	1
Fish kills	1
Changes in benthic fauna	1
Polluted fish traps	1
Deteriorated fishery	1

Source: After Rosenthal (1985).

conversion of the animals, on feeding regimes, and on the composition of the feed (Phillips et al., 1993); part of this nutrient addition is due to excretion and part to unused food. Very dense concentrations of salmonids in European culture operations and of yellowtail fish (*Seriola quinqueradiata*) kept in net enclosures in Japan's Seto Inland Sea are also known to have given rise to enrichment of the underlying sea bottom through siltation and to a reduction in dissolved oxygen in ambient waters. Mollusk cultures in Hiroshima Bay and in the bays of Galicia in Spain have been reported to me to be in tenous equilibrium between an excess of nutrients in the water, which causes oxygen loss, and a lack of the fertilizing elements needed to produce the dense plankton necessary for successful mass culture. Finally, Jaw-Kai Wang of the University of Hawaii, Department of Agricultural Engineering, relates that eutrophication in some waters along the Chinese coast may well be due in part to shrimp pond and fishpond effluents (personal communication, 1995). He also notes that it is difficult to separate out the several causes of enrichment and to pinpoint instances where aquaculture is the sole culprit. Often, though, concern over these alterations of the environment is expressed by the aquaculture farmers themselves, who see in them a threatened reduction in the carrying capacities of their installations rather than a diminished quality of the environment in which they operate.

PARASITES AND DISEASES

Concentration of individual organisms, be they fish, birds, or even humans, in any one location inevitably enhances the opportunity for proliferation of their parasites and diseases. The transfer of worm or crustacean "hangers on" (such as flatworms) on gills, or crustacean lice (*Argulus*) that bury under scales, is made easy, and whole populations in a culture can become infested. Bacteria and viruses have an even easier time of it in water; the use of therapeutic agents has therefore grown by leaps and bounds, with less regulation and official control than in terrestrial situations (Meyer, 1994). Austin (1993) lists 28 compounds that control microbial fish diseases; many are antibiotics, but there are also sulfa preparations and other (some metallic) chemicals among them. These compounds are used far more in freshwater culture and in temperate climes than in tropical mariculture, and there is usually greater concern for their residues in the flesh or on the body of the fish than in the effluent. With the relatively scarce information about biocide persistence in the latter, however, and in view of the rapid genesis of bacterial resistance to antibiotics (Brown et al., 1994), the matter of effect on the effluent deserves attention. Comparable development of chemical control has been attempted for crustacean or worm parasites in fish and mollusks (e.g., the blister worm *Polydora*, a mollusk parasite). These therapeutic agents, mostly organic chemicals containing a heavy metal, are not species-specific but have general biocidal effects on invertebrates. The environmental danger of their excessive use is obvious. Aquacultural farmers, especially with little technical preparation, tend to resort to increased dosages with untoward effects in the environment. Spokespersons for aquacultural research therefore stress that the use of disease-controlling chemicals should be the last and not the first resort in management.

Hardly any chemical treatment has been developed against the viruses of cold-blooded animals, though some vaccines are now used in salmon culture (See Chapter 5). The remedy against a viral outbreak in most culture installation is to eradicate the stock and start again with disease-free animals, as was done with the shrimp *Penaeus stylirostris* in Hawaii. The potential implication for biodiversity here is that affected animals may escape from the cultures into the natural, surrounding environment. However, since viruses are genus- or, more usually, species-specific, this poses no great threat to biodiversity unless the cultured species escaping are also prevalent in the environment.

Still, recent reports on the transfer of shrimp viral outbreaks illustrate certain unanticipated dangers; they concern the survivial of yellow head virus in frozen shrimp and its transfer from Chile and Ecuador to *Penaeus setiferus*, native in certain ponds in Texas where this virus had not been previously noted (G. Pruder, The Oceanic Institute, Waimanalo, Hawaii, personal communication, 1995). The threat such transfer poses very much depends on the density of the hosts in nature. For culture, it is usually serious and demands establishment of disease-free stock, as already mentioned, and use of clean, filtered water.

Aside from the apparent ease of transfer of an aquatic animal virus from one location to another, cause for concern has been expressed about human viral diseases in connection with certain aquacultural practices (Scholtissek and Naylor, 1988). The avian influenza can enter pigs via ducks and chickens, and the human influenza virus can also reside in pigs. Apparently pigs can serve as "vessels" in which the reassortment of the genetic material of avian and mammalian viral strains can take place. The result is new virulence in human influenza; the famed Asian flu pandemics earlier in the 20th century is suspected to have been the result of just such genetic reassortments.

Integrated aquaculture-agriculture in China and Southeast Asia brings together pigs, domesticated birds, and the people who cultivate them around fishponds. Human health hazards, therefore, exist by bringing together two animal reservoirs of genetic materials of influenza, and vigilance has been advocated (Scholtissek and Naylor, 1988) in view of increasing density of land occupancy and intensification of aquacultural practices. But pigs and ducks have been together on Chinese farms for a long time, and it must be said with all caution that genetic changes in the human influenza virus have not (yet) been connected clearly and unequivocally to Asian farm fishponds.

SITING PROBLEMS

Since rearing aquatic organisms requires that the cultivator have control over all or most phases of the life cycle of the cultivars, these are usually confined in one way or another. Ponds, cages, pens, and fish shelters—or, in the case of mollusks and seaweeds, attachment devices—are the instruments for such control or confinement. Water depth and quality and bottom configuration are constraints to their establishment.

In the most rapidly growing variant of tropical aquatic animal husbandry, namely shrimp culture, brackish-water coastal ponds are and will likely remain the bases of production. Necessary attention to viral contamination may, however, require greater control of water quality than such ponds can afford. Costs notwithstanding, tanks and raceways may therefore appear, if not prevail, in shrimp culture of the future.

Shrimp culture can well be represented by shrimp production, which is given in metric tons (MT) per kilometer of coastline and which in the last decades has increased in all Asian nations, except in Taiwan where saturation was reached around 1990 (Csavas, 1993). Shrimp production ranged from 0.5 MT in Malaysia to 37.1 MT in Thailand; in China it is 10.8 MT, and its regional average is 2.1 MT.

It has been recognized that growth of shrimp culture has destroyed substantial mangrove areas, in spite of the fact that mangrove soils are not well suited for this enterprise; they are acid, require frequent liming, and their productivity is difficult to maintain. But entitlement conditions where land may be cheap, free, easily obtained from government sources, or under disuse or poorly applied government control has made for ease in the transformation of mangroves into shrimp farms. In addition, canals, dikes, and the area near the water's edge facilitate the necessary water exchange in ponds even while the latter contribute to altering the hydrologic charac-

TABLE 4.2. Tons of Marine Fish Landed and their Wholesale Value, Malaysia, 1979–1982 (in millions of Malaysian $)

	Year				
	1978	1979	1980	1981	1982
Peninsular west coast[a]					
Quantity	411	432	493	433	434
Value	966	810	779	808	976
Peninsular east coast					
Quantity	154	139	130	216	133
Value	327	178	199	407	258

[a]The west coast still has well-developed mangrove forests; on the east coast they were largely destroyed.
Source: Field and Dartnell (1987).

teristics of the area. Perhaps more important than affording new space, destruction of mangroves also reduces coastal fisheries (Table 4.2). It may be doubted that shrimp culture can make up for these losses in socioeconomic terms.

The impact of shrimp culture on mangrove resources is difficult to interpret because no reliable long-term statistics exist and because various estimates have been done at different times. What is more, mangroves have been removed for all sorts of purposes; in some cases these areas were only later transformed into shrimp ponds. In China, for instance, agricultural land gained from the sea by dyking was later used for shrimp farming because of the increased return per unit of surface, and in Ecuador, a prominent shrimp-producing country, only 24% of the culture areas are believed to have replaced mangrove associations. But whatever the original nature of the culture sites, they are usually close to human settlements, a condition that makes for greater visibility of mangrove destruction. Rough estimates of mangrove transformations into shrimp ponds in a few countries (Table 4.3) can be gleaned from Phillips et al. (1993).

Aquacultural installations in the water, like cages or pens, restrict rather than preempt space and do so to a substantially smaller extent than ponds. Cages and pens may pose obstacles to fishing, recreation, or navigation, and their placement is

TABLE 4.3. Mangrove Reduction into Shrimp Ponds

Approximate Mangrove Areas (ha)		Reduced to by Shrimp-fish ponds	Approximate % reduction
Philippines	448,000	110,000	75 (1988)
Thailand	287,300	249,000	13 (1979)
Indonesia	4,000,000	3,800,000	5 (?)

Source: Data from Phillips et al. (1993).

usually governed and managed by these considerations. But, as mentioned earlier, they can introducte foreign species to an area or sometimes create very large, local accumulations of biomass in small spaces; as a result, they tend to have secondary influences on resident and surrounding biotas, such as eutrophication (see Table 4.1) or the introduction of parasites and diseases. Also, escaped cultivars can affect surrounding wild populations by interbreeding.

SPECIES INTRODUCTIONS

The majority of introductions of new aquatic species worldwide has been made for the purposes of aquaculture (Eldredge, 1994). Some were destined for recreation, as with the spread of trout throughout the former British empire. A later, almost circumglobal, thrust of introductions was the spread of tilapias (*Oreochromis*, mostly *mossambicus*) and the silver carp (*Hypothalmichthys molitrix*), now the most abundant monocultured species in the world. Eldredge (1994) lists over 40 examples of such transportations within the Pacific Islands alone, where the fish were originally brought from one of the continents (though not from South America). Another example of worldwide species translocations is shrimp of genus the *Penaeus*.

Other species introductions were also made with the hope of creating new fisheries, as was the case with the release of the taape (*Lutjanus kasmira*) in 1965 in Hawaii. Most culture-oriented introductions were made to produce food, but nonedible commodities were also involved, as with pearl oysters or colloid-producing algae. Many introductions are so recent, ecologically speaking, that influences on resident biotas were difficult to establish. Also, attempts have not been or could not be made to separate influences of introduced species from those caused by environmental variables such as variations in temperature or nutrient levels, whether these were occurring naturally or due to human activities.

Whatever the economic benefits of aquaculture may be, in any one location there are potential hazards to the natural environment and its biota connected with it, as already mentioned. The unintentional transfer of diseases and parasites is one that may concern cultivars and resident species alike. Giant clams are now being reintroduced into various Pacific Island environments, and with them came some unexpected "riders." The snail predator *Cymatium muricinum* is thought to have been taken to American Samoa along with the giant clam *Tridacna derasa* from Palau, and the ectoparasite *Tathrella iradalei* (a pyramidellid snail) came to Guam from Palau in the same manner (Eldredge, 1994). Only time will tell whether these accidental additions to a local species complement pose a wider threat or whether they remain the worry only of the aquaculturists, as may well be the case.

Accidental introduction by escape from a culture site should also be considered; it is rarer than deliberate releases, which are made with the only-too-understandable intent to improve on nature. Examples abound on land and in the water of such escapes. The establishment of the shrimp *Penaeus merguensis* in a Fijian estuary is an example of an escape from culture in the Pacific Island region, and concerns have been voiced over escapes from salmon culture sites in Europe (Eldredge, 1994).

The instances of moving aquatic animals and plant species into new environments are too numerous to assess. Many but not all have been for the purpose of establishing culture operations, but some have been intended to enhance fisheries. It is important to note that these introductions also start with culture, and when hatchery-produced juveniles are released, they are the product of mating a few individuals with the aim of maximum survival of their larvae and fry. They are therefore less varied genetically than the wild stocks they are intended to bolster. When they then interbreed with individuals of those stocks, a genetically weakened progeny is likely to result (Cautadella and Crossetti, 1993). Concerns of this kind were voiced for several races of salmon in northern Europe and some fish in African lakes, but they could also apply to other locations. Hilo Bay in Hawaii, for instance, was very successfully restocked with mullet produced at the Oceanic Institute on Oahu (Leber, 1995). Not only was this a feat of aquaculture inasmuch as mullet are proverbially difficult to spawn, but the release operation—air-pressured ejection of the small fish through a fire hose—was so successful that 35% of later recaptured tagged fish came from these releases. However improved the sport fishery was, the chances are great for eventually creating a smaller natural gene pool when these fish mate with the resident stock, and thus for reducing the species' ability to adapt to environmental stresses or changes.

Replenishment of locally depleted mullet populations is only one example of attempts to increase numbers in threatened or endangered marine species. The Office of Technology Assessment of the U.S. Congress in its ongoing examination of the role of aquaculture in the United States has commissioned a paper entitled "Aquaculture of Endangered and Threatened Species and Ecosystem management" (Rudloe et al., 1994). Aside from various freshwater species, these authors list three species of marine turtles and two of sturgeons as threatened or endangered, and they describe more or less successful attempts at species restoration. Also treated are attempts at restoring stocks of redfish (*Scianops ocellatus*), Queen conch (*Strombus gigas*), giant clams (*Tridacnidae*), and spiny lobster (*Panilurus argus*); Rudloe and colleagues call these latter species "imperiled." They most likely are in many parts of their ranges, and it goes without saying that the same concerns of reduction in genetic variability apply to all of them, as apply to the mullet of Hawaii, to salmon, and to any other species into which cultured species mates are introduced by intent or by chance if they escape from pens or cages. Concerns about the relationship of aquaculture and restocking to genetic variability have developed together with advances in cytogenetics for the former and with systems ecology in regard to the latter. As past, present, and likely future patterns of environmental change are investigated, adaptive resilience to changes in temperature, water chemistry, and the like appear as increasingly important properties of animal populations, races, and stocks. Loss of genetic variability reduces this resilience and adaptability. Thus, recommendations have been formulated and codes of practice developed for fish introductions and transfers (Turner, 1987; Schramm and Piper, 1995). Yet, because breeding for desirable aquacultural properties, such as growth, uniformity of size, flesh quality, and the like, reduces variability and rewards genetic uniformity, sterile cultivars are the answer here (see also Chapter 5).

ECOSYSTEM RESTORATION

In addition to attempts at restoring the numbers in single species by means of aquacultural techniques, the restoration of degraded coastal habitats—in other words, ecosystem restoration—may become an important, broadly gauged conservation attempt in the 21st century. Commercial fish stocks are certainly not going to recover by themselves while wetlands, especially mangrove stands and seagrass beds, are being lost to various kinds of direct human occupancy the world over. That they are prime nursery grounds for many commercial species has been amply documented (see, e.g., Table 4.3), and it is not unreasonable to assume that attempts to conserve and restore them will assume increasing importance. Like all ecosystems they have a plant base, and artificial propagation of all aquatic plants (not only algae) may someday be listed under the category of aquaculture (Bardach et al., 1972).

Two important points should be noted in this context, however. First, propagation of aquatic plants is technically less difficult than propagation of aquatic animals with their various life-history stages, food requirements, and so forth, and so the conservation of aquatic plants with the intent of recolonizing them from existing stands is cheaper, broadly speaking. Second, if *de novo* plantings are attempted, even in locations that may have been previously colonized, the aquaculturists must have an understanding of prevailing climate, hydrology, geology, and water regimes. Knowledge of these factors and of the social conditions that lead to changes in coastal ecosystems is often lacking. Still, mangroves can be and have been replanted (Hamilton and Snedakar, 1984), leading to new stands—though not to new, fully functioning, complex mangrove ecosystems that would enhance an offshore fishery (see Table 4.3). Sea grasses, among other types of aquatic vegetation, are far more difficult to culture and transplant (Rudloe et al., 1994).

CONCLUSION

Aquaculture in the coastal zone and in fresh waters is an important part of food production especially in Asia (Table 4.4). Its various activities are undertaken for profit, even most of those done on a subsistence scale. It is just one of many uses which society makes of coastal amenities and renewable natural resources. Its processes alter the local natural environment even while they may have further reaching influences like those on the gene pools of resident natural populations.

Each aquacultural installation has its peculiar properties, but like so many other coastal-zone activities, aquacultural effects are not dissimilar and so add up; in certain stretches of the coastal zone, aquaculture can cause notable changes. Pond siting in mangroves is an example here, as are other uses that push beyond carrying capacity of water, as prevailed, for example, until very recently in Laguna Lake, a coastal freshwater lake near Manila. As these changes add up regionally, they become equally potent global change agents to those phenomena expected to act on the global commons, such as global atmospheric warming and a sea-level rise.

TABLE 4.4. Importance of Fish and Aquaculture in Asian Nutrition

Country	1987 Population (millions)	Total Fish Production (MMT)	Fish as % of Animal Protein	Aquaculture Production (MMT)	Aquaculture as % of Fish Production	Aquaculture as % of Animal Protein
Bangladesh	107.0	0.75	52.2	0.13	17	8.87
China	1,062.0	7.05	33.0	3.20	45	14.85
India	800.3	2.86	15.1	1.18	41	6.19
Indonesia	174.9	2.37	67.9	0.30	13	8.83
Malaysia	16.1	0.79	41.6	0.06	9	3.74
Nepal	17.8	0.0006	0.1	0.0004	67	0.07
Philippines	61.5	2.05	56.7	0.50	24	13.61
Sri Lanka	16.3	0.22	39.4	0.036	16	6.30
Thailand	53.6	2.23	51.5	0.14	6	3.09

Source: Bardach (1990).

The most important overall influences on the environment and thus on biodiversity are those of site competition and the effects of effluents. The spread of parasites and diseases and of genetic effects due to aquaculture has so far been less prominent, even if only because these effects are as yet very poorly documented. But they will spread together with aquaculture in general, since demand for aquacultural products will increase as the world population grows. More aquaculture can only have more deleterious effects on the environment if its development is undertaken without attention to ecological and socioeconomic consequences. Present concerns about sustainability do, however, suggest that aquaculture can now develop in a more responsible manner compared with the last decades of its rapid worldwide growth.

Finally, and of relevance not only to islands, is the question of whether proper care and controls are in place to deal with plant and animal introductions. Island ecosystems can be more fragile than those of continental landmasses and coasts, especially as concerns endemic species (Pimm et al., 1995). This is especially true where robust newcomers are introduced anywhere, either for economic considerations or adventitiously, as happened with the tilapia *Oreochromis mossambicus*. Such newcomers can displace native species or depress their populations and can carry diseases to them; diseases may also come by different routes, as was the case with a virus carried by frozen shrimp (see above). Yet introductions for aquaculture are generally justified as a means for economic development. Thus decisions about whether to permit entry, and under what conditions or with what safeguards one might possibly do so, require authorities to balance carefully the potential dangers to the native biota with the improvements in income and earning power for a (not infrequently) needy part of the population.

The plant and animal quarantine rules, regulations, and institutions usually concern themselves with introductions. The efficiency and rigor with which they operate can easily determine the status of island biodiversity on a longer time scale. An internationally recognized, well-working pattern governing introductions prevails in Hawaii, where the Department of Agriculture relies on advice from two tiers of experts in matters of plant and animal introductions. The first tier consists of committees of taxonomists while the second, higher one, also consisting of scientists, but those, such as zoo directors, with administrative experience, is also asked to balance considerations for ecology and biodiversity with those of a social and economic nature.

If these questions are answered satisfactorily and the conditions they pose are accounted for, aquaculture can contribute to responsible—some call it sustainable—development, albeit with the realization that change in the environment will occur but that the severity of these changes can be minimized and their direction guided. In fact, the tenets of sustainable development are just such guided change and the search for multidisciplinary management measures necessary to bring it about.

REFERENCES

Arrow, K., B. Bolin, R. Costanza, P. Dasgupta, C. Folke, C. S. Holling, B. O. Jensson, S. Levin, R. G. Mäler, C. Penings, and D. Pimentel. 1995. Economic growth, carrying capacity and the environment. Science, 268 (April 28): 520–521.

Austin, B. 1993. Environmental issues in the control of bacterial diseases of farmed fish. pp. 237–247, In: Pullin, H. Rosenthal, and J. L. Maclean (Eds.). Environment and aquaculture in developing countries. Conference Proceedings 31. ICLARM. Manila, Philippines.

Bardach, J. E. 1990. Aquaculture and food: opportunities and constraints. Trans. R. Soc. Can. 6th Ser., Vol. I: 371–386.

Bardach, J., J. H. Ryther, and W. O. McLarney. 1972. Aquaculture, Wiley-Interscience. 868 p.

Bloom, D. E. 1995. International public opinion on the environment. Science, 269 (July 21): 355–358.

Brown, L. R., H. Kane, and D. M. Roodman. 1994. *Vital signs, 1994*. W. W. Norton, New York.

Cautadella, S., and D. Crossetti. 1993. Aquaculture and conservation of genetic diversity. pp. 60–73, In: Environment and aquaculture in developing countries. Conference Proceedings 31. ICLARM, Manila, Philippines.

Csavas, I. 1993. Aquaculture development and environmental issues in the developing countries. Asia, 74–100. Conference Proceedings 31. ICLARM, Manila, Philippines.

Eldredge, L. G. 1994. The introduction of commercially significant aquatic organisms to the Pacific Islands. SPREP Reports and Studies Series 78. South Pacific Commission, Noumea, New Caledonia.

Eldredge, L. G., and S. E. Miller. 1995. How many species are there in Hawaii? Bishop Museum Occasional Papers Bishop Museum Press, Honolulu, Hawaii. 18 p.

Eldredge, L. G., J. E. Maragos, and P. L. Holthus. 1996. Marine/coastal biodiversity in the Tropical Island Pacific region. Vol. II: Population, development and conservation priorities. East-West Center, Honolulu, Hawaii.

Field, L. D., and A. D. Dartnell (Eds.). 1987. Mangrove ecosystems of Asia and the Pacific: status, exploitation and management. Australian Institute of Marine Sciences, Townsville, Queensland. 320 p.

Gowen, R. J., and H. Rosenthal. 1993. The environmental consequences of intensive coastal aquaculture in developed countries: what lessons can be learned. pp. 102–115, In: Conference Proceedings 31. ICLARM, Manila, Philippines.

Hamilton, L. and S. C. Snedackar (Eds.). 1984. Handbook for mangrove area management. East-West Center, Honolulu, Hawaii.

Hane, D., and T. Buckley. 1994. *Tridacna derasa* introduction in American Samoa, In: L. G. Eldredge (Ed.). The introduction of commercially significant aquatic organisms to the Pacific Islands. SPREP Reports and Studies Series 78. South Pacific Commission, Noumea, New Caledonia.

Laws, E. A. 1994. Aquatic pollution, John Wiley & Sons, New York.

Leber, K. M. 1995. Significance of fish size-at-release on enhancement of striped mullet fishery in Hawaii. J. World Aquacult. Soc., 26(2): 143–153.

Maclean, J. L. 1993. Developing on country aquaculture and harmful algae blooms. pp. 252–284, In: R. S. V. Pullin, H. Rosenthal, and J. L. Maclean (Eds.). Environment and aquaculture in development countries. Conference Proceedings 31. ICLARM, Manila, Philippines.

Meyer, F. P. 1994. Health and disease management in aquaculture: science, technology and the federal rose. Report prepared for the Office of Technology Assessment, U.S. Congress, Food and Renewable Resources Program, Washington, D.C.

National Research Council, National Academy of Sciences, National Academy of Engineering, Institute of Modern Medicine. 1984. Reducing disaster' toll: the U.S. decade for natural disaster reduction. National Academy Press, Washington, D.C.

Phillips, M. J., C. Kweilin, M. C. M. Beveridge. 1993. Shrimp culture and the environment: lessons from the world's most rapidly expanding warm water aquaculture sector. pp. 171–197, In: R. S. V. Pullin, H. Rosenthal, and J. L. Maclean (Eds.). Conference Proceedings 31. ICLARM, Manila, Philippines.

Pimm, S. L., G. J. Russell, J. L. Gittleman, and T. M. Brooks. 1995. The future of biodiversity. Science, 269(21): 347–354.

Rosenthal, H. 1985. Constraints and perspectives in aquaculture development. Geojournal 10(3): 305–324.

Rudloe, T., T. Madei, and A. Rudloe. 1994. Aquaculture of endangered and Threatened species and ecosystem management. Paper prepared for the Office of Technology Assessment of the U.S. Congress, Washington, D.C.

Scholtissek, Ch., and E. Naylor. 1988. Fishfarming and influenza, pandemics. Nature, 331: 215.

Schramm, H. L., and R. G. Piper (Eds.). 1995. Uses and effects of cultured fishes in aquatic ecosystems. American Fisheries Society Symposium 15, Bethesda, Md.

Turner, G. E. (Ed.). 1987. Codes of practice and manual procedure for consideration of introductions and transfers of marine and fresh water organisms. Prepared for Working Group of Introduction and Transfers of Marine Organisms of ICES and by the Working Party on Introduction of EIFAC. ICES Document F:35A. Copenhagen. 42 p.

5 The Role of Biotechnology in Sustainable Aquaculture

EDWARD M. DONALDSON

INTRODUCTION

With the world population projected to continue its inexorable growth and double between the years 1980 and 2025 to a total of over 8 billion, the production of aquatic foods will have to increase from the present 100 million metric tons (MMT) to 165 MMT in the year 2025 to maintain existing availability per capita (New, 1991). As stated in other chapters, the wild harvest through hunting and gathering appears to have reached its upper limit, an obvious area with potential for growth is aquaculture. In order to have any possibility of meeting the goal of 165 MMT, it will be necessary to develop and apply with great vigor biotechnologies that will facilitate the economic culture of aquatic organisms.

The transition from hunting and gathering to aquaculture is already evident. Thus in Norway, the depletion of wild salmon stocks combined with the rapid growth in aquaculture has led to the farmed production of Atlantic salmon exceeding 100 times the wild catch. In Canada the commercial fishery for wild Atlantic salmon is now closed, and the production of farmed Atlantic and Pacific salmon now exceeds in value the total commercial catch of wild and hatchery-enhanced salmon.

There are several stages in the transition from wild harvest to intensive aquaculture. Intermediate stages can include ocean ranching in which cultivation occurs during only part of the life cycle, extensive culture under natural environmental conditions, intensive culture under natural environmental conditions, and intensive culture under controlled environmental conditions. With each stage of intensification there is increasing opportunity for the application of biotechnology.

The development of a candidate species for aquaculture involves a number of distinct stages, which have been described earlier (Donaldson, 1988). Sustainability depends on successful reproduction and completion of the life cycle in captivity. In most cultivated species the life cycle has been completed by research and development work on broodstock maturation, induced spawning, incubation, larval rearing, and metamorphosis. However, there are notable exceptions such as the culture of eel

Sustainable Aquaculture, Edited by John E. Bardach
ISBN 0-471-14829-6

(*Anguilla* sp.) in Japan and black tiger prawn (*Penaeus monodon*) in Thailand, where large-scale culture is still dependent on the capture of wild larvae and broodstock, respectively, for further progress.

> The completion of the life cycle is essential in any species where later stage juveniles are in limited supply and is also essential before any genetic selection can be undertaken. Subsequent stages involve the development of nutritionally complete and economic diets for each life history stage, the development of diagnostic techniques, disease treatments and vaccines for diseases particular to the species, the refinement of production strategies to meet market requirements and genetic improvement through selective breeding (Donaldson, 1988).

This chapter reviews a number of ways in which biotechnology has played, is playing, and will play a specific role in the development of sustainable aquaculture. Many of the research areas described are relevant to the culture of both finfish and shellfish.

There are several similarities, but also several differences, between the development of biotechnologies for animal and poultry husbandry and that for aquaculture. First of all, the key research areas are similar in both and include reproduction, nutrition, genetics, and health. A major difference lies in the length of the production cycle. Typically the cultivated avian species have a production cycle measured in weeks, while mammalian species are produced in months, and finfish and shellfish have production cycles that extend from months in some tropical species to years in most cold-water species. Furthermore, while only four main avian species—chicken (*Gallus gallus*), turkey (*Meleagris gallopavo*), duck (*Anas platyrhynchos*), and goose (*Anser anser*)—and five main mammalian species—cattle (*Bos taurus*) and (*Bos indicus*), swine (*Sus scrofa*), sheep (*Ovis aries*), and goat (*Capra hircus*)—account for the vast majority of the worldwide production, aquaculture presents a much more diversified array of species. There are many different species of finfish (teleosts), shellfish (crustaceans and mollusks) and macroalgae that are grown worldwide. This can largely be explained by the fact that the avian and mammalian species are air-breathing and homeothermic (i.e., their body temperatures remain held relatively constant independent of the surroundings). Air presents a relatively constant environment (in terms of percentage oxygen, ratio of gases, and so on) on earth, while water is extremely variable in composition. Furthermore, most aquatic species are poikilothermic (i.e., with a body temperature dependent on the surrounding environment), and each species is adapted to thrive over a particular, narrow temperature range. Thus the combination of many different water-quality parameters, including pH, flow rate, turbidity, oxygen concentration, and especially salinity and temperature has led to the culture of many different aquatic species, each adapted to a particular combination of water-quality parameters.

Many areas of biological science are playing a key role in the development of sustainable aquaculture. This chapter interprets biotechnology in the broad sense to include all modern biological technologies that are critical to the successful development of aquaculture, including such areas as broodstock maturation, induced ovulation and spermiation (sperm release), chromosome set manipulation, sex control, incubation and larval rearing, development and metamorphosis, nutrition, improvement of growth and food conversion, fish health, genetics, stock identification, gene banks, and transgenic fish. (Transgenic fish are fish whose genome contains DNA inserted from another species or rearranged from within the same species.)

Many of these biotechnologies focus on the earliest life-history stages of the cultured organism, since it is at this stage that manipulations such as gamete cryopreservation, transgenesis, chromosome manipulation, cloning, and control of sex differentiation take place.

INDUCED SPAWNING

By facilitating the use of domestic broodstock to replace the capture of wild broodstock or fry, biotechnologies for the induction of ovulation and spermiation have already made a significant contribution to the development of sustainable aquaculture. These technologies have evolved from the use of fresh pituitary glands through the use of preserved glands and purified fish and mammalian gonadotropin preparations (e.g., human chorionic gonadotropin) to current third-generation procedures that utilize synthetic nonapeptide (nine amino acid) analogs of hypothalamic peptides, the gonadotropin releasing hormones (GnRHa) including luteinizing-hormone-releasing hormone (LHRHa). These analogs are substituted with a suitable D amino acid in position 6, and the terminal glycine is deleted and replaced with an ethylamide group, thus increasing potency and slowing degradation. One analog based on mammalian LHRH that has seen wide use in aquaculture is [D-Ala^6 des Gly^{10}] LHRH ethylamide. Others have been derived from the sequences of piscine and avian GnRHs. The analogs are typically administered by intraperitoneal or intramuscular injection (Donaldson and Hunter, 1983), but they can also be administered orally (Sukamasavin et al., 1992; Breton et al., 1995) or by polymer implant (Solar et al., 1995). In many marine species, administration of GnRHa alone is sufficient to induce ovulation or spermiation, while in some fresh-water species, and especially carps, it is necessary to block the inhibitory influence of dopamine on gonadotropin release by coadministering a dopamine antagonist such as domperidone (Peter et al., 1988). As our understanding of the basic endocrinology of reproduction in aquatic organisms increases, we can expect the development of new induced-spawning technologies for fish and mollusks and the development of strategies for crustaceans.

SEX CONTROL

The production of monosex and sterile stocks for aquaculture is of benefit for both the optimization of production strategies and for the reproductive containment of genetically altered aquatic organisms (Devlin and Donaldson, 1992; Donaldson et al., 1993). In many fish species and some crustacean species, one sex grows faster, matures later, or has a higher value than the other sex. Thus in flatfish (e.g., turbot, *Scophthalmus maximus*; and olive flounder, *Paralichthys olivaceous*) the female grows more quickly than the male, while in the chinook salmon (*Oncorhynchus tschawytscha*) the male matures on average a year earlier than the female, and in tilapia (*Oreochromis* sp.) culture of monosex male populations prevents reproduction during grow-out. As well as enhancing productivity, monosex stocks could also be used for the reproductive containment of genetically modified aquatic organisms (GMAOs) in situations where no wild conspecifics are present (e.g., Atlantic salmon in Pacific rim waters). The production of sterile aquatic organisms also has a dual purpose. Sterilization can be utilized to prevent precocious or seasonal reproduction in production systems (e.g., rainbow trout, *Oncorhynchus mykiss*; Pacific oyster, *Crassostrea gigas*) and can also be used to prevent the reproduction of GMAOs. Thus it is expected that there would be a requirement for the sterilization of transgenic aquatic organisms used in aquaculture (see below).

Sex control can be achieved by endocrine manipulation during early development (Hunter and Donaldson, 1983; Piferrer and Donaldson, 1993; Pandian and Sheela, 1995), by chromosome set manipulation (Thorgaard, 1983) or by a combination of these. Endocrine manipulation can involve direct masculinization (Piferrer et al., 1993), direct feminization (Piferrer and Donaldson, 1992), or direct sterilization (Piferrer et al., 1994). Where possible, however, direct endocrine sex control is utilized as a step in the production of monosex gametes for the indirect production of monosex stocks that have not received androgen or estrogen treatment (Donaldson, 1996). Recently Y-chromosome-specific DNA probes have been developed to enable sorting fish by genotype regardless of phenotype. This has already greatly facilitated the production of monosex female (XX) stocks of certain salmonids (Devlin et al., 1994b), and it is expected that similar probes will be developed for other economically important species.

Chromosome set manipulation techniques for sex manipulation are usually used in conjunction with endocrine techniques. Thus the production of triploid females (i.e., those containing three sets of chromosomes) results in largely sterile individuals, while in triploid males, testicular development results in the production of aneuploid (having an abnormal number of chromosomes) sperm (Benfey et al., 1986).

The production of monosex female triploid salmon (Donaldson, 1986; Donaldson and Devlin, 1996) and tilapia (Hussain et al., 1995) can be achieved by a variety of routes. Triploidy induction has also been used as a means of limiting reproductive development in oysters thus optimizing product quality (Downing and Allen, 1987). Gynogenesis (i.e., production of a zygote containing only maternal chromosomes) has been utilized in salmonids as a means of producing monosex female (XX) embryos, which can then be masculinized to produce phenotypic males that generate monosex female (X-bearing) sperm. In tilapia the production of YY supermales (Mair et al., 1995) and YY females has been investigated as means to generate monosex male populations.

CLONING

Studies over a decade ago in the zebra fish (*Brachydanio rerio*) demonstrated the feasibility of producing identical (cloned) populations by one cycle of mitotic gynogenesis (diploidy restored by inhibition of the first cell division) followed by one cycle of meiotic gynogenesis (diploidy restored by retention of the second polar body) (Streisinger et al., 1981). In the common carp (*Cyprinus carpio*), androgenic clones (Bongers et al., 1995) and gynogenetic clones (Komen et al., 1991) have been produced for research purposes. Clones have also been produced on an experimental basis in ayu (*Plecoglossus altivelis*; Taniguchi et al., 1988), hirame (*Paralichthys olivaceus*; Yamamoto, 1992), and salmonids (*Oncorhynchus* sp.; H. Onozato, personal communication). It is too early to determine whether the development of cloned populations will remain a research tool or it will be applied in aquaculture (Thorgaard, 1992).

LARVAL REARING: NUTRITIONAL BIOTECHNOLOGY

While some species such as salmonids have been reared on formulated feeds for several decades, many other finfish and shellfish remain dependent on live feeds. There has been considerable research on selection of appropriate species, development of technology for mass culture, optimization of nutritional value (especially polyunsaturated fatty acid content), and the development of feeding strategies. Thus in the development of live feeds for the sablefish (*Anoplopoma fimbria*), the effect of nutrient source on the carbohydrate and fatty acid composition of the rotifer (*Brachionus plicatilis*) has been investigated, and a correlation between larval survival and dietary omega-3 fatty acids, especially eicosapentaenoic acid (EPA), has been demonstrated (Whyte et al., 1994). The importance of omega-3 fatty acids in early rearing has also been investigated in a tropical Pacific species, the dolphin fish (*Coryphaena hippurus*; Ostrowski and Divakaran, 1990). In the rearing of Pacific oyster larvae, the addition of bacterial strain CA2 to xenic (containing naturally occurring microbes) cultures of larvae that were fed the phytoplankton species *Isochrysis galbana* consistently enhanced larval growth (Douillet and Langdon, 1994). It has been proposed that free amino acids in the larvae of marine fish may comprise the main source of energy during embryogenesis and that this may continue after the initiation of exogenous feeding (Fyhn, 1989). If true, this may have significance in the formulation of larval diets.

Recently, cryopreserved oyster trochophores (*Crassostrea gigas*) have been developed as a food source for larval marine species. These can be transported frozen and then thawed as required and fed live to, e.g., larval Nassau grouper (*Epinephelus striatus*; Watanabe et al., 1994). For other species, research continues on the development of formulated diets for early rearing. Key aspects include nutritional value, inclusion of feed attractants, feed type and form (i.e., dry, semimoist, moist, particulate, microencapsulated, paste, flake, gel), and feeding strategy. Microencapsulated diets produced with liposomes and alginate as encapsulating materials have recently been tested in cod (Homme et al., 1994). Aspects of the early rearing of several cold-water marine species including the Atlantic cod (*Gadus morhua*) have been recently reviewed (Tilseth, 1990). Both live and formulated (including microencapsulated) feeds deserve additional research, to reduce cost and improve survival in the mass production of marine species. For several desirable marine species, such as the sablefish (*Anoplopoma fimbria*), the difficulty of larval rearing is the main factor inhibiting their culture.

DEVELOPMENT AND METAMORPHOSIS

Environmental Manipulation

Salmonids undergo a form of metamorphosis when they transform their physiological adaptation from fresh water to seawater in a process known as smolting (Hoar, 1988). It has been shown that the timing of smoltification can be manipulated by photoperiod in both Atlantic (Saunders et al., 1989) and Pacific (Clarke et al., 1989) salmon. This technology is now used to produce smolts for aquaculture much earlier than the natural spring smoltification period. Smolting is accelerated by application of a short photoperiod, followed by a long one.

In the larval abalone (*Haliotis rufescens*), gamma-aminobutyric acid (GABA) mimetic molecules, originating from specific crustose red algae and binding to specific chemosensory receptors, have been shown to trigger a signal transduction cascade that results in metamorphosis and settlement on suitable substrate (Degnan and Morse, 1993). This natural process of metamorphosis can be induced by simple analogs of the exogenous regulatory molecules, and these methods are now in use in the commercial culture of abalone and other mollusks (Morse, 1994).

Endocrine Manipulation

In smolting salmonids, thyroid hormones have long been known to play a regulatory role (Dickhoff and Sullivan, 1987; Bern and Nishioka, 1993). In addition, the pituitary hormones, especially prolactin and growth hormone (somatotropin), are involved in smolting. Recently a dramatic acceleration of smolting was achieved in Pacific salmon presmolts by treatment with recombinant bovine placental lactogen (Devlin et al., 1994a). It is expected that growth hormone or placental lactogen treatments will be used in future to facilitate the transfer of salmonids from fresh to salt water and to optimize posttransfer performance.

In the Japanese flounder (*Paralichthys olivaceus*), thyroid hormones have been shown to play a major role in the transition from the free-swimming to the bottom-living form; furthermore, prolactin and growth hormone gene expression increase during metamorphosis (de Jesus et al., 1993). In the Pacific threadfin (*Polydactylus sexfilis*), treatment of larvae with thyroid hormone and cortisol has improved survival (Brown and Kim, 1995).

To date the endocrine regulation of development and metamorphosis in fish is still in the research and development process. In crustaceans, studies of the endocrine system are leading to the isolation of hormones such as methyl farnesoate that regulate ecdysis (Chang et al., 1993).

GROWTH ACCELERATION

The application of growth acceleration technologies has the potential to shorten the production cycle and improve feed conversion efficiencies (Donaldson et al., 1979) thus reducing waste (Mayer and McLean, 1995). Research has focused on either the administration of either thyroid and steroid hormones (Higgs et al., 1982), or the somatotropins, placental lactogens, prolactins, and related hormones in the growth regulatory system such as somatocrinin, (anti)somatostatin (Mayer et al., 1994), and somatomedin. The recombinant somatotropins (McLean and Donaldson, 1993) and bovine placental lactogen (Devlin et al., 1994a) are the most effective growth stimulants identified to date. For practical use in aquaculture these peptides will be administered in one of three ways: orally as a feed component (McLean et al., 1993b), by polymer implant (McLean et al., 1992), or by sustained release injectable formulation (McLean et al., 1993a). In tilapia (*Oreochromis mossambicus*), dietary administration of the anabolic androgen 17α-methyltestosterone stimulated growth in fresh water- and even more so in seawater-reared fish (Ron et al., 1995). The application of growth acceleration biotechnologies in aquaculture faces regulatory hurdles; however, it offers more flexibility with regard to timing of administration than growth acceleration by transgenesis (see below).

NUTRITION

Research and development on the nutrition of aquatic organisms are of major economic importance, as feed can comprise as much as half of the operating cost of an aquaculture facility (Higgs, 1986). The determination of nutritional requirements at specific life-history stages for the protein, lipid, carbohydrate, mineral, and vitamin content of diets is of key importance in maximizing the efficiency of nutrition in currently cultivated species and in the development of diets for potential aquacultural species.

Information on nutritional requirements facilitates the development of diets for specific life-history stages. Larval or starter diets may contain relatively costly ingredients that stimulate the initiation of feeding, support rapid growth, and promote good survival. Grow-out diets account for the major portion of manufactured diets and must support efficient growth with good conversion efficiency while maintaining fish health. Finishing diets are fed prior to market and are formulated to optimize flesh quality, appearance, and nutritional value at harvest. Broodstock diets are designed to maximize fecundity and gamete quality. The large-scale commercial manufacture of diets for species such as catfish and salmon involves computerized least-cost formulation that integrates knowledge of nutritional requirements with ingredient cost and availability. Diets for aquatic organisms require a high degree of quality control on ingredients and on the manufacturing process. These diets have typically been more complex than those for cultured avian and mammalian species; however, fish diets are now being successfully manufactured with fewer ingredients.

There are two aspects of nutrition that are critical to the long-term sustainability of aquaculture: the need for alternative protein resources, and the need to develop diets that reduce the input of nitrogen and phosphorus into the environment. Biotechnology is also being applied to the development of pigment sources for aquacultural diets and to the optimization of omega-3 fatty acid levels.

Development of Alternative Protein Sources for Aquacultural Diets

Fish meal prepared from wild-caught low-value marine species has been, and still is, the key source of protein in the diets of cultured aquatic organisms (see also Chapter 1). However, the growth of aquaculture is such that the world supply of quality fish meal will be insufficient to meet the demand. North America, for example is a net importer of fish meal, so if aquaculture is to continue to grow, alternative protein sources must be developed. Plant protein provides considerable promise in this regard. Thus processed soybean meal is a key ingredient in catfish diets (Reis et al., 1989), while in tilapia diets, alfalfa leaf protein concentrate (Olvera-Novoa et al., 1990) and canola meal (Higgs et al., 1990) have been used experimentally as dietary protein sources. Recently, complete replacement of fish meal with a specially processed canola protein concentrate was achieved on an experimental basis in the rainbow trout (Higgs et al., 1994). In the blue catfish *(Ictalurus furcatus)*, fish meal has been totally replaced with soybean meal in an experimental trial (Webster et al., 1994). Plant protein in the form of soybean meal and distiller's dried grains with solubles has been used on a trial basis in prawn diets (Tidwell et al., 1993). The utilization of other plant products, including aquatic plants, in aquaculture diets would depend on their nutrient content and cost.

Low-Pollution Diets

Manipulation of the form and content of diets and feeding strategies is facilitating increases in production efficiency and reducing waste output into the receiving environment (Cowey and Cho, 1991; Mayer and McLean, 1995). Thus the development of a soft dry pellet for yellowtail (*Seriola quinqeradiata*) in culture in Japan has improved feed conversion efficiency and reduced excretion of nitrogen and phosphorous. The development of high-energy diets with low carbohydrate and ash contents and optimized protein/energy ratio has improved feed conversion and reduced phosphorus and nitrogen excretion in Atlantic salmon. The administration of somatotropin has been shown to improve conversion efficiency, reduce ammonia excretion, and improve nitrogen retention.

Application of Biotechnology to Pigmentation

The characteristic flesh color of wild salmonids derives from dietary carotenoids (Torrissen et al., 1989). In cultured salmonids it is necessary to administer carotenoid pigments in the diet, which currently account for 15% of the dietary cost. Currently synthetic canthaxanthin and/or astaxanthin are added to the diet throughout the seawater grow-out period for salmon. Astaxanthin is the natural pigment; however, synthetic astaxanthin differs in its isomer content from natural astaxanthin. Currently biotechnology is being applied to pigment production. The yeast (*Phaffia rhodozyma*) produces a natural astaxanthin that is similar to the astaxanthin present in wild salmon; with the development of (*Phaffia*) strains that produce high levels of astaxanthin, it is now feasible to provide natural pigment to salmon via the diet. (Higgs et al., 1995).

Carotenoids are also used to improve the external coloration of the red sea bream (*Pagrus major*) and the pigmentation of cultured prawns (Yamada et al., 1990). In salmonids there is a genetic component to flesh pigmentation (Withler and Beacham, 1994) suggesting that it may be possible to select strains with improved ability to take up and retain astaxanthin.

Optimization of Polyunsaturated Fatty Acids in Fish in Relation to Human Health

There is now a vast literature on the beneficial effects on human health of omega-3 polyunsaturated fatty acids that derive from aquatic organisms (Lands, 1989). The highly unsaturated fatty acids eicosapentaenoic acid (EPA; 20:5*n*-3) and docosahexaenoic acid (DHA; 20:6*n*-3) are the fatty acids of particular interest. In farmed catfish and tilapia the combined concentrations of EPA and DHA have been measured as 2.1 g and 2.3 g, respectively, per 100 g of total fatty acids (Clement and Lovell, 1994). These concentrations are lower than those measured in wild catfish and tilapia (Hearn et al., 1987). It is known that the fatty acid profile of fish reflects the fatty acid profile of the diet. It should therefore be possible to adjust the fatty acid composition of farmed fish by either feeding them a finishing diet that is higher in omega-3 polyunsaturated fatty acids (Dosanjh et al., 1992) or using other strategies (Donaldson et al., 1994), thus providing a form of "designer seafood." In the distant future it may be possible to engineer fish with the innate ability to synthesize highly unsaturated fatty acids.

HEALTH

The development of fish-health biotechnologies is critical to the successful development of sustainable aquaculture. This is especially true for intensive culture under high stocking densities where the stress responses to culture conditions associated with the close proximity of individual organisms can result in the rapid spread of viral, bacterial, and parasitic organisms. Some aspects of fish health where biotechnology is playing a significant role are discussed in the following sections.

Stress Evaluation and Management in Aquaculture

While the basic mechanisms of the stress response in fish have been known for some time (Donaldson, 1981; Mazeaud and Mazeaud, 1981), the role of stress in aquaculture and its causes, effects, and management in relation to overall fish health have been clarified more recently (Barton and Iwama, 1991; Pickering, 1992). Of particular importance is the finding that stress, in particular corticosteroids, cause immunosuppression and increased susceptibility to disease. Thus in an *in vitro* study, cortisol was shown to suppress salmonid lymphocyte responses. (Tripp et al., 1987). It has also been shown that glucans, a form of long-chain polysaccharide, can stimulate nonspecific defense mechanisms in fish (Robertsen et al., 1990; Jeney and Anderson, 1993). In the future it should be possible to minimize stress and its effects in aquaculture by reducing the exposure to stress, understanding the influence of social factors in creating stress (Schreck, 1981), and utilizing the products of biotechnology such as glucans to enhance disease resistance. It may also be possible to improve disease resistance by breeding for a low stress response (Fevolden et al., 1993). A recent publication provides detailed information on a wide range of methods for stress assessment (Adams, 1990).

Disease Diagnostics

For sustainable aquaculture under high-density rearing conditions, the ability to rapidly detect and diagnose disease organisms is of critical importance. This is true both during grow-out and during gamete collection from adult broodstock to reduce or eliminate disease transmission. Biotechnology has made a major contribution to the development of disease-diagnostic tests that are both sensitive and rapid. Thus tests based on fluorescent antibodies and enzyme-linked immunoassays have been compared in the diagnosis of bacterial kidney disease (*Renibacterium salmoninarum*; Olea et al., 1993). The use of nucleic acid probes in aquatic bacteriology with emphasis on the molecular hybridization system has recently been described (Vivares and Guesdon, 1992), and the use of immunoreagents and nucleic acid probes to diagnose diseases in mollusks and crustaceans has been reviewed (Mialhe et al., 1992). The polymerase chain reaction (PCR) has been used to detect bacterial kidney disease DNA in single salmon eggs (Brown et al., 1994), and a PCR test is now available to detect furunculosis (*Aeromonas salmonicida*). U.S. researchers have recently described the use of gene probes to detect four penaeid shrimp viruses—infectious hypodermal and hematopoietic necrosis virus (IHHNV), hepatopancreatic parvovirus (HPV), *Baculovirus penaei* (BP), and monodon baculovirus (MBV)—by dot-blot hybridization, *in situ* hybridization, and PCR (Poulos et al., 1994). It is expected that these new diagnostic techniques will greatly facilitate broodstock screening and monitoring during grow out.

Vaccine Development

In recent years there has been significant progress in the development of vaccines to prevent a wide range of diseases in both finfish and shellfish. Thus in salmonids vaccines are available for vibriosis, yersiniosis, and furunculosis, and their routine use has reduced the frequency of disease outbreaks and the consequent use of antibiotics. However, the development of a vaccine for bacterial kidney disease has proven to be difficult (Evelyn, 1993; Sakai et al., 1993). The development of vaccines is now underway against several salmonid viruses, including viral hemorrhagic septicemia virus (VHSV), infectious pancreatic necrosis virus (IPNV), and infectious hematopoietic necrosis virus (IHNV) (Leong et al., 1994). Vaccines can be administered by immersion, orally, and by automated injection.

GENETICS

The development of domesticated broodstocks with optimized production characteristics is essential to sustained aquaculture. The application of genetics in aquaculture has been the topic of several recent symposia including those in Norway in 1988 (Gjedrem, 1990), China in 1991 (Gall and Chen, 1993), and Canada in 1994 (Doyle et al., 1995). The scientific application of genetic selection combined with improvements in husbandry has resulted in significant improvements in productivity in the culture of aquatic organisms. Selection can be expected to provide major improvements in the economics of culture in cases where species have recently been brought into captivity or been cultured through their complete life cycle. The benefits of selection are not available in partial-life-cycle culture systems that depend on the collection of wild gametes or juveniles.

The production of several aquatic species, including rainbow trout, Atlantic salmon, coho salmon, and oysters, has benefited from selection; however, in these and many other species where selection is possible (e.g., catfish and other salmonid sp.), selection has not been pursued with all the vigor that it deserves. There are significant capital and operating costs associated with a long-term selection program, which are a significant deterrent toward the initiation of such programs. As the selection process progresses in key aquacultured species, we can expect a gradual shift toward organisms that are increasingly domesticated and that potentially would exhibit lower fitness in the wild.

In future we can expect the integration of DNA technology into selection programs. Thus Herbinger et al. (1995) used five hypervariable single-locus microsatellite probes to determine the parentage of communally reared progeny on a rainbow trout farm. This information is being used to reduce the crossing of related broodstock (thus minimizing inbreeding) and to optimize the choice of broodfish in the next spawning season. If this approach proves to be advantageous, the use of DNA probes to match offspring to parents may see widespread use in situations where individual-family rearing units are not available to support a selection program.

STOCK IDENTIFICATION USING DNA TECHNOLOGY

Stocks

There has been considerable research activity in the application of DNA technology to stock identification in aquaculture. Thus in chinook salmon the B2-2 probe has been used to distinguish between stocks raised at different hatcheries and between farmed stocks and local wild stocks, thus facilitating the identification of escaped farm fish (Stevens et al., 1993). Random amplified polymorphic DNA (RAPD) markers have been used in efforts to distinguish between different populations of the sea scallop (*Placopecten magellanicus*) (Patwary et al., 1994). Analyses of length polymorphisms in mitochondrial DNA (mt DNA) restriction fragments have been used to distinguish between stocks of prawn (*Penaeus monodon*; Benzie et al., 1993), grey mullet (*Mugil cephalus*; Crosetti et al., 1993), rainbow trout (*Oncorhynchus mykiss*; Ferguson et al., 1993), and many other aquacultural species.

Individual Offspring to Parents

Many types of DNA fingerprinting probes have been used as genetic markers to identify individuals and family groups, and could also be used to label broodstock and offspring for ownership identity. Thus Harris et al. (1991) have used Jeffreys's multilocus probes to produce DNA fingerprints in the Nile tilapia (*Oreochromis niloticus*), and Herbinger et al. (1995) have used hypervariable single-locus microsatellite probes to relate parents to offspring in rainbow trout. In chinook salmon, individual male and female parents have been identified for particular offspring. (Stevens et al., 1993).

TRANSGENIC FISH

Transgenic aquatic organisms can be defined as having within their genome novel DNA constructs introduced through the use of recombinant DNA technology. Research on the development of transgenic fish with improved production characteristics has increased rapidly over the last several years (Warmbrodt and Stone, 1993; Woodwark et al., 1994). There is currently active research and development underway in many countries including the United States, Canada, the United Kingdom, France, Norway, Japan, China, Taiwan, and New Zealand.

Research on transgenic fish for aquaculture (for review see Hew and Fletcher, 1992) has advanced more quickly than research on transgenic avian or mammalian species for agricultural food production. This may be partly a result of the technical problems associated with the introduction of DNA into the avian egg and the requirement for intrauterine development of the mammalian embryo. There have also been reports of morphological and physiological changes, including reduced fertility in transgenic domestic animals. To date more than 22 species of fish have been genetically modified using more than 40 different DNA constructs (Woodwark et al., 1994).

Research on transgenic fish for aquaculture is currently focused on three groups of production traits: growth, reproduction, and health and environmental tolerance (Devlin et al., 1994c). Recently studies on growth improvement have been extended to the abalone (*Haliotis rufescens*; Powers et al., 1994).

Successful Methods Development

There are several steps in the generation and testing of transgenic aquatic organisms (McLean and Penman, 1990). The appropriate gene sequence that codes for the desired product must be obtained from a gene library. A suitable promoter must also be obtained and inserted upstream of the coding sequence. The DNA construct may also include specific enhancer sequences. Evidence is accumulating that genomic DNA may be preferable to complementary DNA (cDNA DNA that contains only coding sequences) and that promoters and/or genes from the same or related species or the same vertebrate class may be functionally more effective than genes from unrelated species.

The construct is cloned into a plasmid or phage vector and multiplied in bacterial culture. The cloned construct is then separated from the vector and inserted in linear form into the fertilized egg. In fish, insertion is usually achieved by microinjection into the egg cytoplasm in eggs where hardening of the chorion has been inhibited (Devlin et al., 1995a) or removed (dechorionated eggs). The insertion of DNA constructs into fish eggs has also been attempted by electroporation (Powers et al., 1992) and by sperm-mediated transfer (Sin et al., 1993). Other potential techniques include *biolistics* (high-velocity microprojectiles coated with DNA; Zelenin et al., 1991) and chromosome-mediated gene transfer (Disney et al., 1988). This latter process involves the integration of small surviving chromosome fragments from irradiated sperm during gynogenesis. In the future it may be also possible to utilize embryonic stem cells to introduce transgenes into aquatic organisms.

Integration of the introduced construct into the genome can be determined in fin tissue or erythrocytes by the Southern Blot (Chen et al., 1992) or by PCR (Du et al., 1992a). It is also possible to screen fish for transgenism by incorporating a reporter gene into the construct. Thus, Tamiya et al. (1990) and other researchers have demonstrated luciferase gene expression in transgenic fish.

The DNA construct may be incorporated into the genome at multiple sites and in configurations containing multiple copies. Also, transgenic fish are often *mosaic*, i.e., containing the construct in some cells and not others even within the same tissue.

Expression of the inserted gene can be detected by measurement of messenger RNA (mRNA), by measurement of the expressed protein, or by noting the change in phenotype in the microinjected, founder (G_0) fish (Donaldson and Devlin, 1996). When expression results in a change in phenotype such as growth rate, a wide range of responses can be seen. This range may result from several factors, including location of the inserts in the genome, number of insertion sites, and number and arrangement of copies inserted. Successful transmission of the construct to the F_1 generation depends on the construct being present in the gametes and the degree of mosaicism in the gametes. Transmission to the $G_1(F_1)$ generation is normally less than the theoretical 50% that should occur when nontransgenic gametes are fertilized with gametes from a transgenic fish. Thus in mature transgenic coho salmon, 4 out of 5 carried the transgene in the sperm (Devlin et al., 1995a), and transmission to the G_1 ranged from 2% to 20% (Devlin et al., 1995b).

Having reached the point of successfully producing transgenic fish that have altered phenotypes and that transmit the construct to successive generations, we can expected future research to focus on the following areas: control over site of insertion, control over level of expression, control over site and timing of expression, development of true breeding lines of transgenic organisms, and assessment of phenotypes for potential incorporation into production systems.

Use to Accelerate Growth

The development of transgenic fish with improved growth rates has been the aim of many researchers. It is now clear that this objective has been reached for both G_0 and G_1 fish. Thus, common carp transgenic for a Rous sarcoma virus fused to rainbow trout growth hormone (GH1) transmitted the construct to the G_1 generation at rates of 0% to 42%. F_1 growth rates ranged from below that of the controls to 58% above (Chen et al., 1993). Atlantic salmon that were transgenic for an "all fish" construct consisting of the ocean pout *(Macrozoarces americanus)* antifreeze promoter (AFP) and chinook salmon growth hormone cDNA grew to a size several times that of control fish (Du et al., 1992b). In coho salmon that were transgenic for an "all salmon" construct, the average size of the G_0 generation was 10-fold that of control fish after 1 year (Devlin et al., 1994c).

From these and other results, it is evident that dramatic changes in growth phenotype can be induced by the insertion of appropriate DNA constructs. It is expected that future research will focus on homologous or close-to-homologous constructs. Thus the detailed structure of catfish (Tang et al., 1993) and salmon (Devlin, 1993) somatotropin genes has recently been examined. Extreme growth acceleration in fish can result in morphological deformity, especially in the head region (Devlin et al., 1995b), so it will be necessary to focus on the selection of growth-accelerated phenotypes that are similar in morphology to control fish. Research is also being conducted on the use of putative growth-accelerating constructs containing growth hormone releasing hormone (GHRH) or insulinlike growth-factor (IGF) genes. These studies will contribute to our understanding of growth regulation in fish, but are not expected to yield as good results as those using constructs containing the growth hormone gene, at least in the near future.

Other Uses

For reproductive containment, the production of self-sterilizing transgenic fish would be of considerable value. This may be achieved at the hypothalamic level by manipulation of GnRH expression (Aleström et al., 1992), e.g., by GnRH antisense constructs. At the gonadal level it may be possible to utilize genetic ablation to destroy specific gonadal cell types (Maclean and Penman, 1990). The use of these techniques in aquaculture would require the development of means for inducing sterility in production fish while maintaining the reproductive capacity of the broodstock (Devlin and Donaldson, 1992).

Environmental tolerance is another production characteristic that may be amenable to alteration by transgenesis. Atlantic salmon have been made transgenic for the winter flounder (*Pseudopleuronectus americanus*) antifreeze gene (Fletcher et al., 1992). Antifreeze protein expression has been detected in both G_0 and G_1 salmon; however, expression was not yet at a high-enough level to confer significant freeze resistance. The ability to extend the range of environmental conditions under which a particular species could thrive would have important implications for aquaculture.

Increasing the resistance of fish to disease or parasitic infection could also be of major economic importance. Research that is currently underway on the characterization of heat shock and multiple histocompatibility-complex genes in aquatic species may lead to the development of disease-resistant fish. Also, research on insect-resistant transgenic plants suggests that it may in future be possible to produce fish that are resistant to parasites such as sea lice. Progress in these areas could contribute to sustainability by reducing the need for antibiotic or antiparasitic treatments in aquaculture.

In cultured mammals, transgenesis has been used to produce valuable proteins for human medicine by expressing them in milk (gene pharming). *Gene pharming* is the production and harvesting of a valuable protein such as a human pharmaceutical in a transgenic animal or plant. It is possible that analogous techniques could be utilized to produce valuable proteins in aquatic organisms. It is also possible that transgenic techniques could be developed to enhance the nutritional value of plant protein sources for fish diets, and research is underway to develop plant protein sources that include growth-stimulating peptides or proteins.

To date it has only been possible to produce aquatic organisms that are transgenic for a single gene. In future it may be possible to produce fish that are transgenic for a series of genes that could, for example, facilitate carbohydrate utilization by carnivorous fish (thus reducing feed costs), enable synthesis of omega-3 fatty acids (thus reducing dietary requirements for marine oils), and permit endogenous synthesis of pigments such as astaxanthin (thus obviating the need to add these to the diet). In all reports to date only single promoter/gene combinations have been introduced; however, to influence the above traits a series of genes would have to be successfully introduced.

Regulatory Aspects of Transgenics

There has been much discussion concerning this topic in recent years (Kapuscinski and Hallerman, 1991; Hallerman and Kapuscinski, 1992), and several countries including the United States, Canada (Fisheries and Oceans Canada, 1994), and the United Kingdom (Woodwark, 1984) are in the process of developing guidelines and/or regulations to control research, development, and the application of transgenic technologies in aquaculture. It is generally accepted that for the foreseeable future transgenic broodstock fish would be kept in quarantine and that only sterile transgenics would be permitted for aquaculture production. Exceptions could be situations where the survival of escapees would be impossible (e.g., salmonids grown in Hawaii in cool water pumped from the deep ocean, or tropical fish grown in a northern climate). For other species that are cultured in regions where they do not occur naturally and where there are no indigenous species to hybridize with, monosex culture would be sufficient to prevent reproduction in escaping fish. As rivers, lakes, and oceans and the aquatic organisms within them cross international boundaries, there is a great need for international agreement about the development and implementation of regulations controlling the development of transgenic aquatic organisms. There is no doubt that, with the implementation of appropriate regulations to guide their use, transgenic aquatic organisms will in a decade or so start to revolutionize aquaculture as we know it. The potential productivity gains are so great that it will be difficult to ignore them, especially in regions of the world where current protein production does not meet the demand. In the future it may be possible to contain transgenic aquatic organisms by engineering a requirement for a specific nutrient that is not available in the natural environment and therefore is not available to escapees.

GENE BANKS

Sperm Cryopreservation

Sperm cryopreservation is the freezing of sperm in liquid nitrogen in the presence of cryoprotectant compounds. Once frozen, sperm can be stored for an indefinite period prior to thawing for fertilization of eggs. There has been considerable research on the development of gene banks for aquatic species based on sperm cryopreservation (McAndrew et al., 1993). Gene banks can be used for several purposes (Donaldson, 1988) including (1) the indefinite storage of genetic information from a variety of existing wild strains of a given species as an insurance policy against the loss of a stock or for future incorporation into selective breeding programs or development of genomic libraries; (2) the storage of spermatozoa from valuable selectively bred fish for use in future crosses; (3) the preservation of monosex female or male sperm for generation of monosex stocks; (4) the storage of sperm from one breeding season to the next, where the number of spawners is limited, to increase the effective size of the breeding population and thus reduce inbreeding; (5) the maintenance of control lines in selective breeding programs; (6) the development of interspecific crosses in species with differing spawning seasons.

In Norway, Atlantic salmon sperm from most of the stocks in the country has now been cryopreserved as an environmental protection measure. The Food and Agriculture Organization of the United Nations (FAO) recently held a meeting of experts to discuss the development of gene banks for aquatic organisms (FAO 1993). The development of gene banks could play an important role in both the preservation of biological diversity (Harvey, 1996) and the provision of genetic material for future genetic improvement for aquaculture.

Egg and Embryo Cryopreservation

Egg and embryo cryopreservation involves freezing eggs and embryos in liquid nitrogen in the presence of cryoprotectant compounds for an indefinite period. The cryopreservation of large eggs containing a considerable amount of yolk (e.g., salmonid eggs) presents difficult technical problems as the optimal freezing conditions for the yolk are different from those for the other components. Also, diffusion rates for the cryoprotectants are too slow in large eggs. Quicker progress can be expected in the cryopreservation of the smaller eggs, embryos, and larvae characteristic of nonsalmonids. Thus oyster trochophores have been successfully cryopreserved, and other larvae may be suitable candidates (Rana, 1995).

Role in Aquaculture Development

Gene banks have played a vital role in the development of animal husbandry, and we can expect the same to be true for aquaculture. Cryopreservation has already been used in salmon culture to store valuable monosex sperm from one reproductive season to the next and to develop hybrids between species with differing spawning seasons. Its use may have been slower to develop in, for example, catfish culture because in this species reproduction often occurs in the pond environment while in salmon culture fertilization always takes place in the hatchery. As the development of high-performance selectively bred or transgenic aquatic organisms proceeds, we can expect to see increasing use of gamete cryopreservation as both a means of storage and a means of transport.

INTEGRATION OF SEPARATE BIOTECHNOLOGIES

While many individual biotechnologies can stand alone (e.g., induced ovulation, cryopreservation), in many situations the greatest benefits can be achieved through the pairing of individual technologies. Thus, the sterilization technology that has demonstrated the most promise to date involves combining monosex female production with triploidy induction (Donaldson et al., 1993) This sterilization technology can then be combined with other biotechnologies either to provide reproductive containment, as in the case of transgenic fish (Devlin and Donaldson, 1992; Donaldson et al., 1993) or to optimize production. An example of the latter could be the combination of growth acceleration technologies, such as the placental lactogen or somatotropin administration with triploid induction, so as to accelerate growth and optimize feed conversion without risk of precocious sexual maturation (McLean et al., 1993b).

Further examples of paired technologies include the induction of gynogenesis followed by direct masculinization and the direct masculinization of mixed-sex embryos followed by use of a Y-specific DNA probe to identify genetic sex. Both of these paired technologies result in the efficient production of monosex female sperm in a single generation (Donaldson and Devlin, 1996).

DEVELOPMENT POTENTIAL, RESEARCH, AND REGULATORY PRIORITIES

In comparison with the faltering production and future production uncertainty in many sectors of the wild fishery, sustainable aquaculture appears to have a bright future as a source of human food and other biological products. Key areas for current and future research and development include reproduction, nutrition, health, and genetic improvement. In the area of reproduction, further research is required on the reproduction of new species and improved control over the reproduction of existing species. In nutrition, key research areas include (1) the further development of alternative protein sources, especially the utilization of plant protein; (2) strategies to reduce feed costs; (3) dietary strategies to maximize product quality; and (4) development of dietary means to administer pharmaceuticals and vaccines that optimize health and performance. In health, research priorities for biotechnology lie especially in the areas of diagnosis and prevention, but also in the treatment of disease. Genetic improvement depends on research and development on (1) the application of biotechnology (e.g., DNA fingerprinting) to traditional selection, (2) chromosome set manipulation in its various aspects, and (3) long-term programs to develop improved strains through transgenesis.

Regulatory priorities that stem from the great mobility of many aquatic organisms include the development of and international agreement on (1) guidelines for developing and using genetically modified aquatic organisms (GMAOs), including transgenic organisms in aquaculture; and (2) guidelines to protect native stocks from the impact of the escape of exotic species.

CONCLUSIONS

Biotechnology is already playing a major role in the development of sustainable aquaculture. We can expect the role of biotechnology to become increasingly important. The trend will continue to be driven by market forces, including international competition and consumer preferences, and environmental concerns. The role of biotechnology varies depending on what the stage of development is for the production systems for a particular species. In new species the focus is on using biotechnology to complete the life cycle in culture, thus securing independence from wild seed stock and enabling the application of genetics. For species that have recently entered the marketplace, the focus of biotechnology is on solving production problems such as minimizing disease by developing vaccines. Finally, in mature species where production is stable, the emphasis is on lowering production costs, adding value, and ensuring quality and consistency. If nations are able to apply biotechnology to the enhancement of aquaculture production, as some developed and developing nations are beginning to do (e.g., Cuba and China), aquaculture may possibly relieve the pressure slightly on at least some wild stocks and thus facilitate their continued survival.

REFERENCES

Adams, S. M. 1990. Biological indicators of stress in fish. American Fisheries Symposium 8. Bethesda, Md. 191 p.

Aleström, P., K. Klungland, G. Kisen, and ØO. Andersen. 1992. Fish gonadotropin releasing hormone gene and molecular approaches for control of sexual maturation: Development of a transgenic fish model. Mol. Mar. Biol. Biotech., 1: 376–379.

Barton, B. A., and G. K. Iwama. 1991. Physiological changes in fish from stress in aquaculture with emphasis on the response and effects of corticosteroids. Annu. Rev. Fish. Dis., 1: 3–26.

Benfey, T. J., I. I. Solar, G. DeJong, and E. M. Donaldson. 1986. Flow cytometric confirmation of aneuploidy in sperm from triploid rainbow trout. Trans. Am. Fish. Soc., 115: 838–840.

Benzie, J. A. H., E. Ballment, and S. Frusher. 1993. Genetic structure of *Penaeus monodon* in Australia: concordant results from mtDNA and allozymes. Aquaculture, 111: 89–93.

Bern, H. A., and R. S. Nishioka. 1993. Aspects of salmonid endocrinology: the known and the unknown. Bull. Fac. Fish. Hokkaido Univ., 44(2): 55–67.

Bongers, A. B. J., J. B. Abarca, B. Z. Doulabi, E. H. Eding, J. Komen, and C. J. J. Richter. 1995. Maternal influence on development of androgenetic clones of common carp, *Cyprinus carpio* L. Aquaculture, 137: 139–147.

Breton, B., I. Roelants, T. Mikolajczyk, P. Epler, and F. Ollevier. 1995. Induced spawning in teleost fish after oral administration of GnRH-A. pp. 102–104, In: F. W. Goetz and P. Thomas (Eds.). Proceedings of the Fifth Symposium on Reproductive Physiology of Fish, July 2–8. University of Texas at Austin, Austin.

Brown, C. L., and B. G. Kim. 1995. Combined application of cortisol and triiodothyronine in the culture of larval marine finfish. Aquaculture, 135: 79–86.

Brown, L. L., G. K. Iwama, T. P. T. Evelyn, W. S. Nelson, and R. P. Levine. 1994. The use of polymerase chain reaction (PCR) to detect DNA from *Renibacterium salmoninarum* within individual salmon eggs. Dis. Aquat. Org., 18: 165–171.

Chang, E. S., S. L. Fitzsimmons, and M. J. Bruce. 1993. Crustacean growth: stimulatory and inhibitory regulation. pp. 514–519, In: W. Fenical, M. Greenberg, H. O. Halvorsen, and J. C. Hunter-Cevera (Eds.) International Marine Biotechnology Conference, October 13–16, 1991, Baltimore, MD.

Chen, T. T., C. M. Lin, R. A. Dunham, and D. A. Powers. 1992. Integration, expression and inheritance of foreign fish growth hormone gene in transgenic fish. pp. 164–175, In: C. L. Hew and G. L. Fletcher (Eds.). Transgenic fish. World Scientific, Singapore.

Chen, T. T., K. Kight, C. M. Lin, D. A. Powers, M. Hayat, N. Chatakondi, A. C. Ramboux, P. L. Duncan, and R. A. Dunham. 1993. Expression and inheritance of RSVLTR-rtGH1 complementary DNA in the transgenic common carp, *Cyprinus carpio*. Mol. Mar. Biol. Biotech., 2(2): 88–95.

Clarke, W. C., J. E. Shelbourne, T. Ogasawara, and T. Hirano. 1989. The effect of initial day length on growth, seawater adaptability and plasma growth hormone levels in under yearling coho, chinook and chum salmon. Aquaculture, 82: 51–62.

Clement, S., and R. T. Lovell. 1994. Comparison of processing yield and nutrient composition of cultured Nile tilapia (*Oreochromis niloticus*) and channel catfish (*Ictalurus punctatus*). Aquaculture, 119: 299–310.

Cowey, C. B., and C. Y. Cho (Eds.). 1991. Nutritional strategies and aquaculture waste. Proceedings of the First International Symposium on Nutritional Strategies in Management of Aquaculture Waste. University of Guelph, Guelph, Ontario, Canada. 275 p.

Crosetti, D., J. C. Avise, F. Placidi, A. R. Rossi, and L. Sola. 1993. Geographic variability in the grey mullet *Mugil cephalus*: preliminary results of mtDNA and chromosome analyses. Aquaculture, 111: 95–101.

Degnan, B. M., and D. E. Morse. 1993. Identification of eight homeobox-containing transcripts expressed during larval development and at metamorphosis in the gastropod mollusc *Haliotis rufescens*. Mol. Mar. Biol. Biotech., 2: 1–9.

de Jesus, E. G., T. Hirano, and Y. Inui. 1993. Flounder metamorphosis: its regulation by various hormones. Fish Physiol. Biochem. 11: 323–328.

Devlin, R. H. 1993. Sequence of sockeye salmon type 1 and 2 growth hormone genes and the relationship of rainbow trout with Atlantic and Pacific salmon. Can. J. Fish. Aquat. Sci., 50: 1738–1748.

Devlin, R. H., and E. M. Donaldson. 1992. Containment of genetically altered fish with emphasis on salmonids. pp. 229–265, In: C. L. Hew and G. L. Fletcher (eds.). Transgenic fish. World Scientific, Singapore.

Devlin, R. H., J. C. Byatt, E. McLean, T. Y. Yesaki, G. G. Krivi, E. G. Jaworski, and E. M. Donaldson. 1994a. Bovine placental lactogen is a potent stimulator of growth in coho salmon, and displays strong binding to hepatic receptor sites. Gen. Comp. Endocrinol., 95: 31–41.

Devlin, R. H., B. K. McNeil, I. I. Solar, and E. M. Donaldson. 1994b. A refined PCR-based test for Y-chromasonal DNA allows simple production of all-female strains of Chinook salmon. Aquaculture, 128: 211–220.

Devlin, R. H., T. Y. Yesaki, C. Biagi, E. M. Donaldson, and W.-K. Chan. 1994c. Production and breeding of transgenic salmon. pp. 372–378, In: C. Smith, J. S. Gavora, B. Benkel, J. Chesnais, W. Fairfull, J. P. Gibson, B. W. Kennedy, and E. B. Burnside (Eds.). Selection and quantitative genetics; growth; reproduction; lactation; fish; fiber; meat. Proceedings of the Fifth World Congress on Genetics Applied to Livestock Production, Vol. 19. August 7–12, Guelph, Ontario, Canada.

Devlin, R. H., T. Y. Yesaki, C. Biagi, E. M. Donaldson, P. Swanson, and W.-K. Chan. 1994c. Extraordinary salmon growth. Nature, 371: 209–210.

Devlin, R. H., T. Y. Yesaki, E. M. Donaldson, S. J. Du, and C. L. Hew. 1995a. Production of germline transgenic Pacific salmon with dramatically increased growth performance. Can. J. Fish. Aquat. Sci. 52: 1376–1384.

Devlin, R. H., T. Y. Yesaki, E. M. Donaldson, and C. L. Hew. 1995b. Transmission and phenotypic effects of an antifreeze/GH gene construct in coho salmon (*Oncorhynchus kisutch*). Aquaculture, 137: 161–169.

Dickhoff, W. W., and C. V. Sullivan. 1987. Involvement of the thyroid gland in smoltification with special reference to metabolic and developmental processes. Am. Fish. Soc. Symp., 1: 197–210.

Disney, J. E., K. R. Johnson, D. K. Banks, and G. H. Thorgaard. 1988. Maintenance of foreign gene expression in adult transgenic rainbow trout and their offspring. J. Exp. Zool., 248: 335–344.

Donaldson, E. M. 1981. The pituitary interrenal axis as an indicator of stress in fish. pp. 11–47, In: A. D. Pickering (Ed.). Stress and fish. Academic Press, London.

Donaldson, E. M. 1986. The integrated development and application of controlled reproduction techniques in Pacific salmonid aquaculture. Fish Physiol. Biochem., 2: 9–24.

Donaldson, E. M. 1988. Science and the future of aquaculture. pp. 299–309, In: G. M. Stone and G. Evans (Eds.) Animal reproduction: research and practice. Proceedings of the Aquaculture International Congress and Exposition, September 6–9, Vancouver, B.C. Publ. Aqua. Int. Congr. B.C. Pavilion Corp.

Donaldson, E. M. 1996. Manipulation of reproduction in farmed fish. In: Proceedings of the 13th International Congress on Animal Reproduction (ICAR), June 30–July 4, Sydney. Animal Reproduction Science, 42: 381–392.

Donaldson, E. M., and R. H. Devlin. 1996. Uses of biotechnology to enhance production. Ch. 17, pp. 969–1020 In: W. Pennell and B. Barton (Eds.). Principles of salmonid culture; Elsevier, Netherlands.

Donaldson, E. M., and G. A. Hunter. 1993. Induced final maturation, ovulation, and spermiation in cultured fish. pp. 351–603 In: W. S. Hoar, D. J. Randall, and E. M. Donaldson (Eds.). Fish psychology, Vol. IX, Pt. B. Academic Press, New York.

Donaldson, E. M., U. H. M. Fagerlund, D. A. Higgs, and J. R. McBride. 1979. Hormonal enhancement of growth in fish. pp. 455–597, In: W. S. Hoar, D. J. Randall, and J. R. Brett (Eds.). Fish physiology, Vol. VIII. Academic Press, New York.

Donaldson, E. M., R. H. Devlin, I. I. Solar, and F. Piferrer. 1993. The reproductive containment of genetically altered salmonids. pp. 113–129, In: J. G. Cloud and G. H. Thorgaard (eds.). Genetic conservation of salmonid fishes. NATO ASI Series. Plenum Press, New York.

Donaldson, E. M., R. H. Devlin, D. A. Higgs, and I. M. Price. 1994. Aquaculture biotechnology in Canada including the development of transgenic salmon. pp. 47–57, In: OECD Group of National Experts on Biotechnology, Working Group on Food Safety Meeting on Marine Biotechnology, June 10–13, 1992. Bergen, Norway. OECD Documents, Aquatic Biotechnology and Food Safety, Bergen.

Donaldson, E. M., R. H. Devlin, F. Piferrer, and I. I. Solar. In press. Hormones and sex control in fish with particular emphasis on salmon. Asian fisheries science, 9(1): 1–8.

Dosanjh, B., D. Higgs, G. Deacon, A. Farrell, T. Petryshen, A. Reaney, and R. Brewer. 1992. Influence of dietary lipid composition on growth and muscle lipid composition of chinook salmon in sea water. The 21st Fish Feed and Nutrition Workshop, Oct. 7–9. Davis, Calif. Abstract, p. 10.

Douillet, P. A., and C. J. Langdon. 1994. Use of a probiotic for the culture of larvae of the Pacific oyster (*Crassostrea gigas* Thunberg). Aquaculture, 119: 25–40.

Downing, S. L., and S. K. Allen, Jr. 1987. Induced triploidy in the Pacific oyster, *Crassostrea gigas*: optimal treatments with cytochalasin β depend on temperature. Aquaculture, 61: 1–15.

Doyle, R. W., C. M. Herbinger, M. Ball, and G. A. E. Gall (Eds.). 1995. Genetics in aquaculture, Vol. V. Proceedings of the Fifth International Symposium on Genetics in Aquaculture, June, 1994, Halifax, Canada. Aquaculture 137: 1–358.

Du, S. J., Z. Gong, G. L. Fletcher, M. A. Shears, and C. L. Hew. 1992a. Growth hormone gene transfer in Atlantic salmon: use of fish antifreeze/growth hormone chimeric gene construct. pp. 176–189, In: C. L. Hew and G. L. Fletcher (Eds.). Transgenic fish. World Scientific, Singapore.

Du, S. J., Z. Gong, G. L. Fletcher, M. A. Shears, M. J. King, D. R. Idler, and C. L. Hew. 1992b. Growth enhancement in transgenic Atlantic salmon by the use of an "all-fish" chimeric growth hormone gene construct. Bio/technology, 10: 176–180.

Evelyn, T. P. T. 1993. Bacterial kidney disease—BKD. pp. 177–195, In: V. Inglis, R. J. Roberts, and N. R. Bromage (Eds.). Bacterial diseases of fish. Blackwell Scientific Publications, Oxford.

FAO. 1993. Report of the expert consultation on utilization and conservation of aquatic genetic resources. November 9–13, 1992 Grotta Ferrata, Italy. FAO Fisheries Report # 491, FAO, Rome.

Ferguson, M. M., R. G. Danzmann, and S. K. A. Arndt. 1993. Mitochondrial DNA and allozyme variation in Ontario cultured rainbow trout spawning in different seasons. Aquaculture, 117: 237–259.

Fevolden, S. E., R. Nordmo, T. Refstie, and K. H. Roed. 1993. Disease resistance in Atlantic salmon (*Salmo salar*) selected for high or low responses to stress. Aquaculture, 109: 215–224.

Fisheries and Oceans Canada. 1994. Transgenic aquatic organisms: policy and guidelines for research with, or for rearing in natural aquatic ecosystems in Canada. Draft 3. November 1994. 39 p.

Fletcher, G. L., P. L. Davis, and C. L. Hew. 1992. Genetic engineering of freeze-resistant salmon. pp. 190–208, In: C. L. Hew and G. L. Fletcher (Eds.). Transgenic fish. World Scientific, Singapore.

Fyhn, H. J. 1989. First feeding of marine fish larvae: are free amino acids the source of energy? Aquaculture, 80: 111–120.

Gall, G. A. E., and H. Chen. 1993. Genetics in aquaculture IV. Aquaculture, 111: v–viii + 1–332.

Gjedrem, T. 1990. Genetics in aquaculture III. Aquaculture, 85: v–x + 1–340.

Hallerman, E. M., and A. R. Kapuscinski. 1992. Ecological and regulatory uncertainties associated with transgenic fish. pp. 209–228, In: C. L. Hew and G. L. Fletcher (eds.). Transgenic fish. World Scientific, Singapore.

Harris, A. S., S. Bieger, R. W. Doyle, and J. M. Wright. 1991. DNA fingerprints of tilapia, *Oreochromis niloticus*, and its application to aquaculture genetics. Aquaculture, 92: 157–163.

Harvey, B. J. 1996. Banking fish genetic resources: the art of the possible. pp. 439–446, In: F. di Castri and T. Younes (Eds.). Biodiversity, science and development: towards a new partnership. Publ. CAB International, Wallingford, U.K.

Hearn, T. L., S. A. Sgoutas, J. A. Hearn, and D. S. Sgoutas. 1987. Polyunsaturated fatty acids and fat in fish flesh for selecting species for health benefits. J. Food. Sci., 52: 1209–1210.

Herbinger, C. M., R. W. Doyle, E. R. Pitman, D. Paquet, K. A. Mesa, D. B. Morris, J. M. Wright, D. Cook. 1995. DNA fingerprint based analysis of paternal and maternal effects on offspring growth and survival in communally reared rainbow trout. Aquaculture, 137: 245–256.

Hew, C. L., and G. L. Fletcher (Eds.). 1992. Transgenic fish. World Scientific, Singapore. 274 p.

Higgs, D. A. 1986. Fish nutrition research: a means of increasing the profitability of Pacific salmon farming. Can. Aquacult. Mag., 2(2): 15, 17, 18, 26.

Higgs, D. A., U. H. M. Fagerlund, J. G. Eales, and J. R. McBride. 1982. Application of thyroid and steroid hormones as anabolic agents in fish culture. Comp. Biochem. Physiol. 73B: 143–176.

Higgs, D. A., B. S. Dosanjh, M. Little, R. J. J. Roy, and J. R. McBride. 1990. Potential for including canola products (meal and oil) in diets for *Oreochromis mossambicus* × *O. aureus* hybrids. pp. 301–314, In: Proceedings of the Third International Symposium on Feeding and Nutrition in Fish, Toba, August 28–September 1, 1989.

Higgs, D. A., B. S. Dosanjh, A. F. Prendergast, R. M. Beames, R. W. Hardy, W. Riley, and G. Deacon. 1994. Use of rapeseed/canola in finfish diets. In: Nutrition and Utilization Technology in Aquaculture. AOCS Press, Champaign, Il.

Higgs, D., E. Donaldson, B. Dosanjh, E.-A. Chambers, M. Shamaila, B. Skura, and T. Furukawa. 1995. The case for Phaffia. Northern Aquacult., 11(2): 20–24.

Hoar, W. S. 1988. The physiology of smolting salmonids. pp. 275–343, In: W. S. Hoar and D. J. Randall (Eds.). Fish physiology, Vol. XIB. Academic Press, New York.

Homme, J. M., E. Wathne, and K. Hjelmeland. 1994. Liposome-alginate encapsulation of formulated feed for fish larvae: *in vitro* and *in vivo* studies of digestibility. Third International Marine Biotechnology Conference, August 7–12. Tromso, Norway. Program and Abstracts, p. 62.

Hunter, G. A., and E. M. Donaldson. 1983. Hormonal sex control and its application to fish culture. pp. 223–303, In: W. S. Hoar, D. S. Randall, and E. M. Donaldson (Eds.). Fish physiology, Vol. IXB. Academic Press, New York.

Hussain, M. G., G. P. S. Rao, N. M. Humayun, C. F. Randall, D. J. Penman, D. Kime, N. R. Bromage, J. M. Myers, and B. J. McAndrew. 1995. Aquaculture, 138: 87–97.

Jeney, G., and D. P. Anderson. 1993. Glucan injection or bath exposure given alone or in combination with a bacterin enhance the non-specific defense mechanisms in rainbow trout (*Oncorhynchus mykiss*). Aquaculture, 116: 315–329.

Kapuscinski, A. R., and E. M. Hallerman. 1991. Implications of introduction of transgenic fish into natural ecosystems. Can. J. Fish. Aquat. Sci., 48(Suppl. 1): 99–107.

Komen, J., A. B. J. Bongers, C. J. S. Richter, W. B. Van Muiswinkel, and E. A. Huisman. 1991. Gynogenesis in common carp (*Cyprinus carpio* L.). II. the production of homozygous gynogenetic clones and F_1 hybrids. Aquaculture, 92: 127–142.

Lands, W. E. M. 1989. Fish and human health: a story unfolding. World Aquacult. 20: 59–62.

Leong, J. C., and 10 coauthors. 1994. Viral vaccines for aquaculture. Third International Marine Biotechnology Conference, August 7–12, Tromso, Norway. Program and Abstracts, p. 45.

Maclean, N., and D. Penman. 1990. The application of gene manipulation to aquaculture. Aquaculture, 85: 1–20.

Mair, G. C., J. S. Abulay, J. A. Beardmore, and D. O. F. Skibinski. 1995. Growth performance trials of genetically male tilapia (GMT) derived from YY-males in *Oreochromis miloticus* L: On station comparisons with mixed sex and sex reversed male populations. Aquaculture, 137: 313—322.

Mayer, I., and E. McLean. 1995. Bioengineering and biotechnological strategies for reduced waste aquaculture. Water Sci. Technol., 31(10): 85–102.

Mayer, I., E. McLean, T. J. Kieffer, L. M. Souza, and E. M. Donaldson. 1994. Antisomatostatin-induced growth acceleration in chinook salmon (*Oncorhynchus tshawytscha*). Fish Physiol. Biochem. 13: 295–300.

Mazeaud, M. M., and F. Mazeaud. 1981. Adrenergic responses to stress in fish. pp. 49–75, In: A. D. Pickering (Ed.). Stress and Fish, Academic Press, London.

McAndrew, B. J., K. J. Rana, and D. J. Penman. 1993. Conservation and preservation of genetic variation in aquatic organisms. pp. 295–336, In: J. F. Muir and R. J. Roberts (Eds.). Recent advances in aquaculture, Vol. IV. Blackwell Scientific, Oxford.

McLean, E., and E. M. Donaldson. 1993. The role of somatotropin in growth in poikilotherms. pp. 43–71, In: M. P. Schreibman, C. G. Scanes, and P. K. T. Pang (eds.). The endocrinology of growth, development and metabolism in vertebrates. Academic Press, New York.

McLean, E., E. Teskeredzic, E. M. Donaldson, Z. Teskeredzic, Y. Cha, R. Sittner, and C. G. Pitt. 1992. Accelerated growth of coho salmon *Oncorhynchus kisutch* following sustained release of recombinant porcine somatotropin. Aquaculture, 103: 377–387.

McLean, E., R. H. Devlin, E. M. Donaldson, C. A. Baile, R. J. Collier, and G. G. Krivi. 1993a. Preliminary evaluation of a sustained release somatotropin formulation for use in intensive aquaculture. p. 521, In: World Aquaculture '93, May 26–28, Torremolinos, Spain. European Aquaculture Society Special Publication 19. Oostende, Belgium.

McLean, E., E. M. Donaldson, E. Teskeredzic, and L. M. Souza. 1993b. Growth enhancement following dietary delivery of recombinant porcine somatotropin to diploid & triploid coho salmon (*Oncorhynchus kisutch*). Fish Physiol. Biochem. 11: 363–369.

Mialhe, E., V. Boulo, E. Bachere, D. Hervio, K. Cousin, D. Noel, T. Noel, M. Ohresser, R. M.

le Deuff, B. Despres, and S. Gendreau. 1992. Development of new methodologies for diagnosis of infectious diseases in mollusc and shrimp aquaculture. Aquaculture, 107: 155–164.

Morse, D. E. 1994. Signal molecules, receptors, transducers and genes controlling metamorphosis of marine invertebrate larvae: Application to medicine and aquaculture. Third International Marine Biotechnology Conference, August 7–12, Tromso, Norway. Program and Abstracts, p. 75.

New, M. B. 1991. Turn of the millennium aquaculture. Navigating troubled waters or riding the crest of the wave? World Aquacult. 22(3): 28–49.

Olea, I., D. W. Bruno, and T. S. Hastings. 1993. Detection of *Renibacterium salmoninarium* in naturally infected Atlantic salmon *Salmo salar* L., and rainbow trout *Oncorhynchus mykiss* (Walbaum) using an enzyme-linked immunosorbent assay. Aquaculture, 116: 99–110.

Olvera-Novoa, M. A., G. S. Campos, G. M. Sabido, and C. A. Martinez Palacios. 1990. The use of alfalfa leaf protein concentrates as a protein source in diets for tilapia (*Oreochromis mossambicus*). Aquaculture, 90: 291–302.

Ostrowski, A. C., and S. Divakaran. 1990. Survival and bioconversion of n-3 fatty acids during early development of dolphin (*Coryphaena hippurus*) larvae fed oil-enriched rotifers. Aquaculture, 89: 273–285.

Pandian, T. J., and S. G. Sheela. 1995. Hormonal induction of sex reversal in fish. Aquaculture, 138: 1–22.

Patwary, M. U., E. L. Kenchington, C. J. Bird, E. Zouros, and J. P. Van der Meer. 1994. The use of Random Amplified Polymorphic DNA (RAPD) markers in genetic studies of the sea scallop *Placopecten magellanicus*. Third International Marine Biotechnology Conference, August 7–12 Tromso, Norway. Program and Abstracts, p. 134.

Peter, R. E., H. R. Lin, and G. Van Der Kraak. 1988. Induced ovulation and spawning of cultured freshwater fish in China: Advances in application of GnRH analogues and dopamine antagonists. Aquaculture, 74: 1–10.

Pickering, A. D. 1992. Rainbow trout husbandry: management of the stress response. Aquaculture, 100: 125–139.

Piferrer, F., and E. M. Donaldson. 1992. The comparative effectiveness of the natural and a synthetic estrogen for the direct feminization of chinook salmon (*Oncorhynchus tshawytscha*). Aquaculture, 106: 183–193.

Piferrer, F., and E. M. Donaldson. 1993. Sex control in Pacific salmon. pp. 69–77, In: J. F. Muir and R. J. Roberts (Eds.). Recent advances in aquaculture, Vol. IV. Blackwell Scientific, Oxford.

Piferrer, F., I. J. Baker, and E. M. Donaldson. 1993. Effects of natural, synthetic, aromatizable, and nonaromatizable androgens in inducing male sex differentiation in genotypic female chinook salmon (*Oncorhynchus tshawytscha*). Gen. Comp. Endocrinol., 91: 59–65.

Piferrer, F., M. Carrillo, S. Zanuy, I. I. Solar, and E. M. Donaldson. 1994. Induction of sterility in coho salmon, *Oncorhynchus kisutch*, by androgen immersion before first feeding. Aquaculture, 119: 409–424.

Poulos, B. T., D. V. Lightner, R. M. Redman, H. K. Halsey, J. J. Reddington, J. L. Mari, and J. R. Bonami. 1994. Diagnostic procedures using gene probes for penaeid shrimp viruses. World Aquaculture '94. Book of abstracts, p. 122.

Powers, D. A., T. Cole, K. Creech, T. T. Chen, C. M. Lin, K. Kight, and R. A. Dunham. 1992. Electroporation: a method for transferring genes into the gametes of zebrafish, *Brachydanio rerio*, channel catfish, *Ictalurus punctatus*, and common carp, *Cyprinus carpio*. Mol. Mar. Biol. Biotechnol., 1: 301–309.

Powers, D. A., L. Hereford, T. Cole, and M. Gomez-Chiarri. 1994. Genetic engineering of a fast growing strain of abalone, *Haliotis rufescens*. Third International Marine Biotechnology Conference, August 7–12, Tromso, Norway. Program and Abstracts, p. 70.

Rana, K. 1995. Cryopreservation of aquatic gametes and embryos: recent advances and applications. pp. 85–89, In: F. W. Goetz and P. Thomas (Eds.). Proceedings of the Fifth International Symposium on Reproductive Physiology of Fish, July 2–8. University of Texas at Austin, Austin.

Reis, L. M., E. M. Reutebuch, and R. T. Lovell. 1989. Protein-to-energy ratios in production diets and growth, feed conversion and body composition of channel catfish, *Ictalurus punctatus*. Aquaculture, 77: 21–27.

Robertsen, B., G. Rostad, R. Engstad, and J. Raa. 1990. Enhancement of non-specific disease resistance in Atlantic salmon *Salmo salar* L. by a glucan from *Saccharomyces cereviciae* cell walls. J. Fish Dis., 13: 391–400.

Ron B., S. K. Shimoda, G. K. Iwama, and E. G. Frou. 1995. Relationships among ration, salinity, 17α methyltestosterone and growth in the euryhaline tilapia (*Oreochromis mossambicus*). Aquaculture, 135: 185–193.

Sakai, M., S. Atsuta, and M. Kobayashi. 1993. The immune response of rainbow trout (*Oncorhynchus mykiss*) injected with five *Renibacterium salmoninarum* bacterins. Aquaculture, 113: 11–18.

Saunders, R. L., J. L. Specker, and M. P. Komourdjian. 1989. Effects of photoperiod on growth and smolting in juvenile Atlantic salmon (*Salmo salar*). Aquaculture, 82: 103–126.

Schreck, C. B. 1981. Stress and compensation in teleostean fishes: response to social and physical factors. pp. 277–293, In: A. D. Pickering (Ed.). Stress and fish. Academic Press, London.

Sin, F. Y. T., A. L. Bartley, S. P. Walker, I. L. Sin, J. E. Symonds, L. Hawke, and C. L. Hopkins. 1993. Gene transfer in chinook salmon (*Oncorhynchus tshawytscha*) by electroporating sperm in the presence of pRSV-lacZ DNA. Aquaculture, 117: 57–69.

Solar, I. I., J. Smith, H. M. Dye, D. MacKinlay, Y. Zohar, and E. M. Donaldson. 1995. Induced ovulation of chinook salmon using a GNRHa implant: effect on spawning, egg viability and hormone levels. p. 144, In: F. W. Goetz and P. Thomas (Eds.). Proceedings of the Fifth International Symposium on Reproductive Physiology of Fish, July 2–8. University of Texas at Austin, Austin.

Stevens, T. A., R. E. Withler, S. H. Groh, and T. D. Beacham. 1993. A new multi-locus probe for DNA fingerprinting in chinook salmon (*Oncorhynchus tshawytscha*), and comparisons with a single locus probe. Can. J. Fish. Aquat. Sci., 50: 1559–1567.

Streisinger, G., C. Walker, N. Dower, D. Knauber, and F. Singer. 1981. Production of clones of homozygous diploid zebra fish (*Brachydanio rerio*). Nature (London), 291: 293–296.

Sukumasavin, N., W. Leelapatra, E. McLean, and E. M. Donaldson. 1992. Orally induced spawning of Thai carp (*Puntius gonionotus*, Bleeker) following co-administration of des Gly^{10} (D-Arg^{6}) sGnRH ethylamide and domperidone. J. Fish Biol., 40: 477–479.

Tamiya, E., T. Sugiyama, K. H. Masaki, A. Hirose, T. Ohoshi, and I. Karube. 1990. Spatial imaging of luciferase gene expression in transgenic fish. Nucleic Acids Res. 18: 1072.

Tang, Y., C. M. Lin, T. T. Chen, H. Kawauchi, R. A. Dunham, and D. A. Powers. 1993. Structure of the channel catfish (*Ictalurus punctatus*) growth hormone gene and its evolutionary implications. Mol. Mar. Biol. Biotechnol., 2(4): 198–206.

Taniguchi, N., S. Seki, J. Fukai, and Y. Inada. 1988. Induction of two types of gynogenetic diploids by hydrostatic pressure shock and verification by genetic marker in ayu, *Plecoglossus altivelis*. Nippon Suisan Gakkaishi, 52: 49–53.

Thorgaard, G. H. 1983. Chromosome set manipulation and sex control in fish. pp. 405–434,

In: W. S. Hoar, D. J. Randall, and E. M. Donaldson (Eds.). Fish Physiology, Vol. IXB. Academic Press, New York.

Thorgaard, G. H. 1992. Application of genetic technologies to rainbow trout. Aquaculture, 100: 85–97.

Tidwell, J. H., C. D. Webster, D. H. Yancey, and L. R. D'Abramo. 1993. Partial and total replacement of fish meal with soybean meal and distillers' by-products in diets for pond culture of the fresh water prawn (*Macrobrachium rosenbergii*). Aquaculture, 118: 119–130.

Tilseth, S. 1990. New marine species for cold-water farming. Aquaculture, 85: 235–245.

Torrissen, O. J., R. W. Hardy, and K. D. Shearer. 1989. Pigmentation of salmonids—carotenoid deposition and metabolism. CRC Crit. Rev. Aquat. Sci., 1: 209–225.

Tripp, R. A., A. G. Maule, C. B. Schreck, and S. L. Kaattari. 1987. Cortisol mediated suppression of salmonid lymphocyte responses *in vitro*. Dev. Comp. Immunol., 11: 565–567.

Vivares, C. P., and J.-L. Guesdon. 1992. Nucleic acid probes in aquatic bacteriology. Aquaculture, 107: 147–154.

Warmbrodt, R. D., and V. Stone. 1993. Transgenic fish research: A bibliography. Bibliographies and Literature of Agriculture 117. National Agricultural Library, Beltsville, Md. 48 p.

Watanabe, W. O., C.-S. Lee, S. C. Ellis, E. P. Ellis, W. D. Head, C. D. Kelly, G. Miyamoto, K. Liu, and J. Ginoza. 1994. Experimental culture of larval Nassau grouper (*Epinephalus striatus*): the effects of temperature on egg and yolksac stages and of prey quality on survival at first feeding. World Aquaculture '94. Book of Abstracts, p. 287. World Aquaculture Society, Baton Rouge, La.

Webster, C. D., L. Goodgame-Tiu, and J. H. Tidwell. 1994. Effect of totally replacing fish meal with soybean meal and supplemental methionine on growth of blue catfish (*Ictalurus furcatus*). World Aquaculture '94. Book of Abstracts, p. 185. World Aquaculture Society, Baton Rouge, La.

Whyte, J. N. C., W. C. Clarke, N. G. Ginther, J. O. T. Jensen, and L. D. Townsend. 1994. Influence of composition of *Brachionus plicatilis* and Artemia on growth of larval sable fish (*Anoplopoma fimbria* Pallas). Aquaculture, 119: 47–61.

Withler, R. E., and T. D. Beacham. 1994. Genetic variation in body weight and flesh colour of the coho salmon (*Oncorhynchus kisutch*) in British Columbia. Aquaculture, 119: 135–148.

Woodwark, M., D. Penman, and B. McAndrew. 1994. Genetically modified fish populations. U.K. Department of Environment. 87 p.

Yamada, S., Y. Tanaka, M. Sameshima, and Y. Ito. 1990. Pigmentation of prawn (*Penaeus japonicus*) with carotenoids 1. Effect of dietary astaxanthin, β carotene and canthaxanthin on pigmentation. Aquaculture, 87: 323–330.

Yamamoto, E. 1992. Application of gynogenesis and triploidy in hirame (*Paralichthys olivaceus*) breeding. Fish Gen. 18: 13–23 (In Japanese).

Zelenin, A. V., A. A. Alimov, V. A. Barminstev, A. O. Beniumov, I. A. Zelenina, A. M. Krasnov, and V. A. Kolesnikov. 1991. The delivery of foreign genes into fertilized fish eggs using high velocity microprojectiles. FEBS Lett. 287: 118–120.

6 Economic Decision Making in Sustainable Aquacultural Development

YUNG C. SHANG AND CLEM A. TISDELL

INTRODUCTION

Aquaculture is a fast-growing industry with an average annual growth rate of about 12% during the past decade. Total production in 1993 was 22.6 million metric tons (MMT), including seaweeds (FAO, 1995a). Asia was the leading region in aquaculture, with 85.8% of global production, followed by Europe (7.3%), North America (3.5%), South America (1.5%), the former USSR (1.1%), Oceania (0.4%), and Africa (0.4%).

We can add further to these and earlier statistics: Aquaculture produced about 23% of the total world food fish supply in 1993. The percentage was significantly higher by major commodity groups: 94% of seaweeds, 76% of mollusks, 72% of freshwater fish, 23% of crustaceans, and 1.2% of marine fish supply. Of the total aquaculture production, fish (predominantly freshwater) accounted for 49.5%, seaweeds 27.7%, mollusks 18.2%, crustaceans 4.1%, and others 0.5%. About 47% of aquaculture production came from inland waters and 53% from marine waters. If seaweeds are excluded, the importance of inland waters rises substantially (66%).

There are significant differences between the major commodity groups in terms of their environmental impact. Since seaweeds consume dissolved nutrients and produce oxygen, their production is regarded as the most environmentally compatible practice of aquaculture. Culturing filter-feeder mollusks is also considered environmentally beneficial because they can remove nitrogen and phosphorus from water. However, the amount they can remove is insignificant compared with the organic load to the environment under intensive culture (Csavas, 1994). For example, overuse of traditional oyster culture ground in Korea and Japan has led to self-pollution. In addition, mollusk culture is especially susceptible to environmental changes and water pollution; siltation and sedimentation in the coastal areas can have a harmful effect on mollusk culture. Inland aquaculture produces primarily

Sustainable Aquaculture, Edited by John E. Bardach
ISBN 0-471-14829-6

finfishes; and its development is limited mainly by the scarcity of suitable land and water due to the competition with other economic activities, and by self-pollution. Most of the environmental troubles are caused by the uncontrolled development of coastal aquaculture, especially shrimp farming. Overinvestment in shrimp culture and its attendant environmental degradation has caused the collapse of the shrimp industry in Taiwan and China. Marine fish culture has also had ecological problems due to increasing domestic and industrial pollution, self-pollution, and red tide, which results in massive fish kills. The impacts of aquaculture on the environment and of the environment on aquaculture have attracted the attention of proponents of sustainable development.

Sustainability is a fascinating subject and has emerged in recent years as an important dimension of aquaculture and other aspects of natural resources management and development. There are many definitions and interpretations of sustainability, some of which have been mentioned in Chapter 1. These definitions can be summarized as follows: (1) sustainable aquaculture is usually viewed as a farming system that is in harmony with other economic activities that use natural recourses. (2) It should produce a reasonable and relatively stable net income or benefit to both producers and society, compared with other economic alternatives, by using the same natural resources on a long-term basis without degrading the environment; therefore, the way natural resources are used in aquaculture should not leave future generations worse off than the current generation. (3) Its development has to be balanced among production, marketing, and other supporting services (including legal measures) in order to be sustainable.

RELATIONSHIP AMONG BIOTECHNICAL, ENVIRONMENTAL, AND ECONOMIC FACTORS IN SUSTAINABLE DEVELOPMENT

A sustainable aquaculture system has to be biotechnically feasible, environmentally sound, and socioeconomically viable. These three aspects are interrelated (Figure 6.1). This section briefly reviews their interrelationships, even though some of them are also discussed elsewhere in the book.

Bio-Technical Factors

There are many biotechnical factors affecting the environmental compatibility and socioeconomic viability of aquaculture: the species selected (herbivores, omnivores, or carnivores) and the culture system adopted (extensive, semi-intensive, or intensive; monoculture, polyculture, or integrated culture) are the important ones, as shown in Figure 6.2. (These are also influenced by socioeconomic considerations that will be discussed later.)

Although categories and classifications of aquaculture have been described in Chapter 1 in the section "Sustainability and Patterns of Aquaculture," it is necessary to repeat them briefly here to show their relationship with environmental and economic factors. In traditional culture systems, herbivorous species, and to some

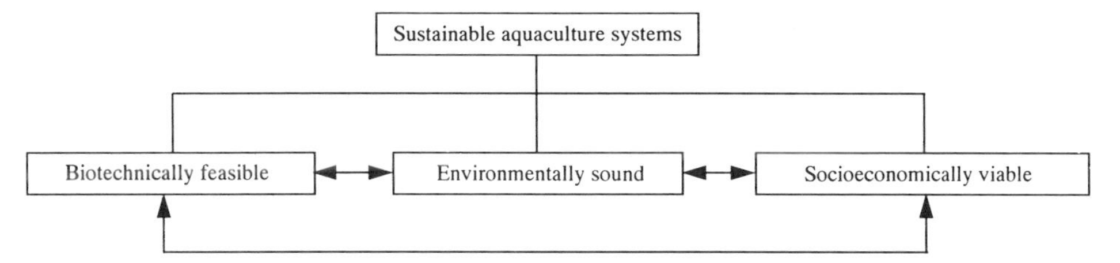

Figure 6.1. Sustainable aquaculture system.

extent omnivores, can rely on natural protein or low-protein food while carnivorous species need high-protein food. The higher the protein level in the feed, the higher the cost of production and the higher the possibility of water pollution under intensive culture. Of aquacultured finfish produced, carnivorous species (e.g., salmonids, catfish, yellowtails, eels, and seabasses) in 1993 accounted for about 15%, and noncarnivorous species (e.g., carps, tilapias, and milkfish) for 85% of the weight.

Extensive culture systems usually have relatively low stocking densities, require little or no supplementary feeding or fertilizing, and normally produce insignificant loading of nutrients or organic matter onto the ecosystem. In contrast, intensive

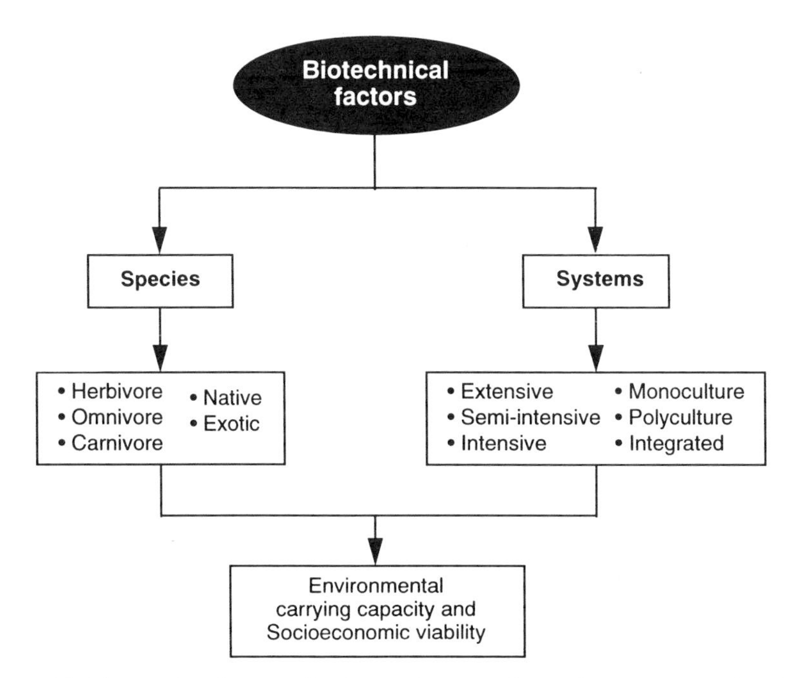

Figure 6.2. Biotechnical factors affecting environmental carrying capacity and socioeconomic viability.

culture systems are characterized by relatively high stocking densities and high levels of input, such as feed and fertilizers, and chemicals and drugs, which normally increase the nutrient and organic matter load on the ecosystem. Intensive culture systems may create considerable pollution if not properly managed. The cost of pollution abatement often limits the enterprise's commercial viability.

Polyculture and integrated farming are usually ecologically more sound than monoculture due to more efficient use of waste and energy on the farm. (These points will be discussed later in the chapter.)

Environmental Factors

Environmental factors affecting sustainable aquacultural development can be viewed from two levels: farm and society (Figure 6.3). At a farm level, fish wastes and unconsumed feed are decomposed in ponds or beneath cages. These wastes often result in high concentrations of nitrogen, phosphorus, and organic matter in the culture facilities (see also Chapter 4). Farm productivity usually increases with culture intensity, but it will eventually decline after a certain level of intensity due to crowding, deteriorated water quality, diseases, and so forth, resulting in reduced growth and high mortality. All of these effects compromise economic viability and sustainability.

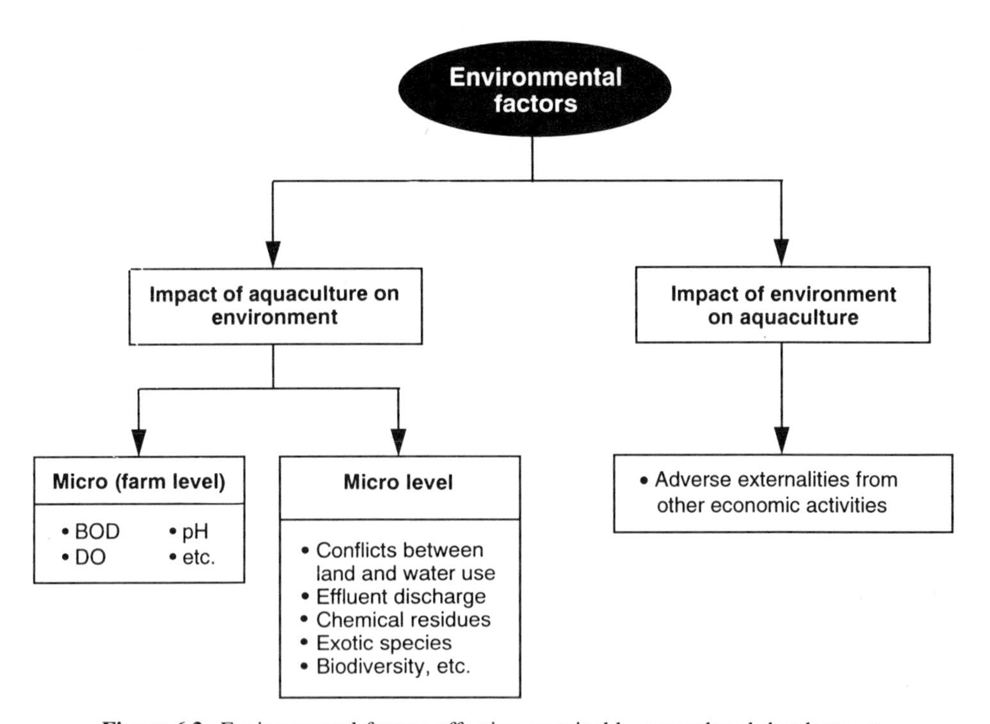

Figure 6.3. Environmental factors affecting sustainable aquacultural development.

From a societal viewpoint, aquaculture utilizes natural resources and is an integral part of the environment. Aquaculture imposes a variety of impacts on the external environment, some positive and some negative. These impacts usually result from conflicts over land and water use, effluent discharge, chemical residues, exotic species culture, and so on, all of which will be discussed later. The expansion of aquaculture and the trend toward intensification have caused concerns about the carrying capacity of farms and, more broadly, of the environment.

On the other hand, aquaculture is often a victim of adverse externalities resulting from other economic activities. Water pollutants such as heavy metals, chemical wastes, and pesticides from industry, agriculture, and domestic sewage can pose serious environmental threats to aquaculture. Without social controls, these adverse impacts may reduce aggregate production and threaten the social welfare. In planning for sustainable aquacultural development, these adverse externalities should be taken into consideration; in part they can be internalized, and in part eliminated by external and internal remedial actions.

Socioeconomic Factors

Socioeconomic factors affecting sustainable aquacultural development include social acceptability and economic viability (Figure 6.4). To be sustainable, the species cultured and the culture systems adopted should be socially acceptable based on local conditions. (For example, tilapia is both culturally accepted and unaccepted—accepted in Asia but generally not esteemed in Europe and North America; others like shrimp are universally liked even though they are luxury species.) There should also be no conflicts with existing or potential resource use practices. Otherwise, there would be limited potential for development. More discussions on the social issues can be found in another chapter of this book.

Economic viability can be also viewed from two levels: farm (micro) and society (macro). The economic viability of a sustainable commercial fish farm depends on how well its productivity is sustained and on market factors reflected in prices paid for inputs and prices obtained for its products. In a social context, the impact of an aquaculture industry cannot be assessed in isolation from the rest of the economy. An aquaculture industry may be sustainable in terms of its productivity or profitability, but this may only be attained at the expense of other economic activities (e.g., shrimp ponds in healthy mangrove areas) or because of pollution from aquaculture. The overall value of production may therefore be reduced and the potential income available to future generations might also be lowered due to natural resource depletion or deterioration (Tisdell, 1995). On both levels, resources have to be used and coordinated in a way to produce the best sustainable net income/benefit over time without degrading the environment. A farm or an industry that is not profitable cannot sustain itself for long economically, no matter how ecologically sound it may be. Likewise, a farm or an industry that is not ecologically sound cannot sustain itself economically no matter how productive or profitable it may be in the short run. (More discussion on economic viability will be found in the following sections.)

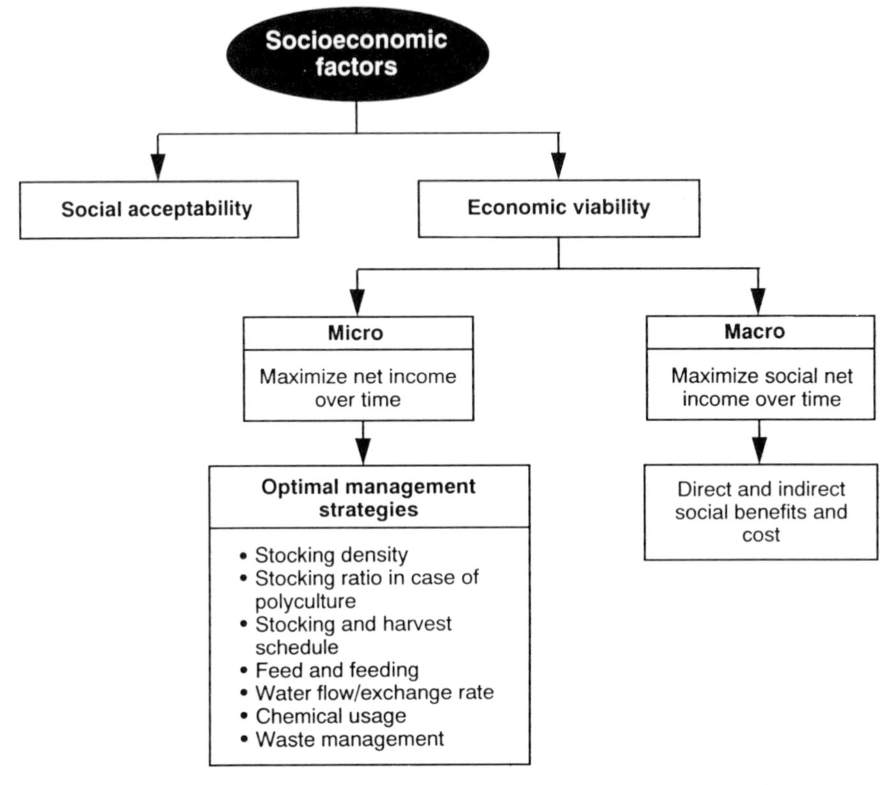

Figure 6.4. Socioeconomic factors affecting sustainable aquacultural development.

ECONOMIC DECISION MAKING IN SUSTAINABLE AQUACULTURAL DEVELOPMENT

Like economic viability, the economic decision making in sustainable aquacultural development can be discussed on the micro (farm) level and the macro (societal) level.

Micro (Farm) Level

At the farm level, the major decisions are what to produce and how to produce it. The species selected for production (a staple and/or high-value species) should conform to the following criteria: be fast-growing, be adapted to the local conditions, be technically feasible, have seed and feed available at a reasonable costs, be socially acceptable (see earlier remark on tilapia), and have a potential market for generating a reasonable profit without degrading the environment. The producer should constantly evaluate his or her decision as to what to produce, basing the decision on future market trends, input and output prices, and improvements in culture technologies.

Many fish farmers often seek to maximize their short-term gain at the expense of the environment. The "rape and run" practice in shrimp farming, where ponds in mangrove areas are farmed intensively and quickly abandoned, as has been observed in Thailand and the Philippines, is a good example. This type of farming system violates the economic criterion of sustainability as mentioned earlier, and therefore, is unsustainable. The objective of a sustainable aquaculture farm should be to maximize its sustainable net income in the long run. Farm net income is simply the difference between total farm revenues and total farm operating costs. Thus, increasing revenues and reducing costs are the major means of increasing net income. This is biotechnical and economic decision making on how to combine inputs to produce a satisfactory if not optimal output mix.

Farm revenue is affected by the level of production and farm price. Increasing farm productivity is one way to increase farm revenue. Farm productivity depends mainly on two sets of factors: (1) stocking rate, survival rate, and growth rate, which are in turn affected by biotechnical factors such as rate of feeding and fertilization, whether the culture is mono- or polyculture, stocking and harvesting strategies, and so on, and (2) environmental factors such as water quality (water temperature, dissolved oxygen, pH levels, and so forth), chemical inputs to treat diseases and predators, and so on. The farm price of an aquacultured product is mainly affected by size, form, and quality of the product in competitive seasonal markets (Shang, 1981). The best combination of inputs and the most suitable culture intensity and/or system is mainly determined by the resources available on the farm and their costs, and the farmer's management ability.

The costs of production, on the other hand, relate to the level of inputs, the prices of inputs, the culture systems, the sizes of operation, the waste treatment as well as the institutional factors such as costs of credit, marketing, land lease, and the like. Neither the costs of waste treatment and pollution prevention, nor the taxes on discharging effluents are usually included in conventional financial analysis, but they are important cost items for a sustainable operation.

The cost of feed is probably the most important cost item for a relatively intensive aquaculture (especially for carnivores), and in many cases feed composes more than 50% of the total operating cost. Cost of feed per unit of fish produced depends primarily on the conversion ratio of feed to fish and the unit price of feed. Therefore, the cost of feed can be reduced by improving the conversion ratio or by lowering the unit price of feed, or by a combination or substitution of these two factors. (See also Chapter 1, passages on fish meal.) The conversion ratio in turn can be improved by reducing waste and improving the feed formula. The waste can be reduced by feeding the right amount of feed, while the unit cost of feed may be lowered by utilizing locally available materials for feed instead of imported ones. The economic principle of feeding is that it should occur at a level such that any additional cost of feed equals its additional revenue. Overfeeding results in a higher cost of feed per unit of fish produced and also in water pollution.

Another practice to reduce the costs of feed and fertilizer is to use integrated aquaculture-agriculture systems. Farm wastes can be recycled to produce multiple products, and water pollution and feed and fertilizer costs can be reduced. This farming practice diversifies and minimizes risks and uncertainty, and safeguards the environment.

In most of the current economic analysis, water, except the pumping cost, is not included as a cost item in aquaculture production. However, as the quantity of water diminishes and its quality deteriorates in the future, suitable water could become an important cost item for a sustainable operation. Deteriorated water quality causes diseases and low survival rate for intensively managed farms. Better pond management is essential to improve water quality. Cleaning up incoming water in settling ponds and using a closed-culture system, sediment management, and polyculture are some of the means for improving water quality if they are economically feasible. The problem of water-quality control in open coastal waters is far more difficult to solve. (See also Chapter 3 on Offshore Mariculture.)

Diseases have emerged as a major constraint to the sustainable growth of aquaculture. In many intensively managed farms, chemicals and antibiotics are used to protect the stocks. Over time, the pond environment slowly deteriorates due to excessive use and residual accumulation of these toxic elements. There are also adverse effects of these elements on consumers due to the retention of chemicals in aquacultural products. Many of the diseases are linked to environmental deterioration and stress associated with the intensification of aquaculture practices. Disease prevention is more important than treatment. The solution to the problem must deal with fish-farm design, sitting and sustainable farm management. From an economic view point, for example, one practice to reduce the environmental stress and the cost of these drugs is to fallow ponds after each harvest instead of applying drugs. Although fallowing the ponds may reduce the number of crops, this may be justified by the savings on drugs and by the long-term benefits of a sustainable production. The economics of alternative ways of disease control need to be assessed and compared. In the long run, genetic improvement of the cultured species is likely to result in disease-resistant strains, greater tolerance to environmental variation, and faster growth. Improved virus-free fry may also reduce disease problems in the grow-out stage.

Methods of Economic Analysis at the Farm Level. Methods of traditional economic analysis can be used for sustainable aquaculture if they are refined and improved to take into account their applications to specific environments. The following methods are commonly used to analyze the net farm income of alternative farming systems and practices and to select the optimal farm management strategy (Shang, 1990). Some of these methods involve mathematical programming, which may be difficult to do by some fish farmers (especially on small farms). In this case, extension agents should be in a position to perform these analyses.

Budget Analysis. This method focuses on costs and returns associated with a particular farming system or practice, and timing. The budgets can be complete or partial. However, if aquaculture is a component of a farm's activities, whole-farm analysis is more appropriate. In addition, the methods of analysis can range from simple spreadsheet models to mathematical programming and simulation. Linear programming can be used to determine the profit-maximizing combination of scarce resources for producing a particular species, given the environmental constraints, and to examine alternative farming practices for reducing negative effects.

Bioeconomic-Environmental Modeling. An aquaculture production system operates like a dynamic and probabilistic process depending on many biotechnical, environmental, and economic elements as mentioned earlier. Sustainable and profitable operation can be achieved only through better understanding of these relevant elements and of their interrelationships in the entire production process. This is not an easy task for any producer without the assistance of a proper tool. A system analysis that can integrate all of these important elements and help the producer to locate the optimal course of action will contribute to the success of the operation. To develop such an integrated analysis, the collaborative efforts of biologists, ecologists, and economists are needed. More discussions on this topic will be found in Chapter 7.

Macro (Societal) Level

Sustainable aquacultural development depends on the rational use of environmental and other natural resources. This can be done by improving project planning and management. At the societal/community level, the major decision making about sustainable aquacultural development involves the objectives of rational development, species for culture, culture systems and techniques, and environmental policy measures.

Objectives of Aquaculture Development. The first step in project selection is to determine the objective(s) of development. Aquacultural development strategies and projects should conform to the objectives of national development, which is ultimately to improve human welfare. From a societal and community level, the management objective of sustainable aquaculture is to maximize sustainable net social benefit. The specific objectives of aquacultural development are summarized below.

Human Food. Aquaculture production increased rapidly from 6.9 MMT in 1984 to 16.3 MMT 1993 (excluding seaweeds). Its contribution to the world fish food supply increased from 12% in 1984 to 23% in 1993. In many developing countries, people derive as much as 50% or more of their daily animal protein requirement from fish (Shang, 1992a). Fish consumption is projected to increase in the future due to increasing population and income and to changing consumption patterns for health reasons. With a forecast world population of about 7.032 billion in 2010, the Food and Agriculture Organization of the United Nations (FAO) projected that 91 MMT of food fish would be required based on present per capita fish consumption of 13 kg/y alone (FAO, 1995b). This requirement implies an increase of 19 MMT of food fish over the 1993 level of supply (72.3 MMT).

The above estimated demand for fish assumed a population increase in which the present level of consumption is maintained. However, when personal income increases, people will consume more fish. Projection of income growth is more difficult than that of population. The projected growth rate of the world's real gross domestic product for the next decade is about 4.7% (Shang, 1992b). If we assume the same percentage increase in real personal income and an income elasticity of 0.14 in the demand for fresh fish (Cheng and Cappos, 1988), the projected demand for fish due to income increase by the year 2010 is 11% more than the 1993 consumption level; this is about 8 MMT, assuming no significant change in fish prices. The combined increase in demand for fish due to population and income is estimated at about 27 MMT by the year 2010, which is about a 37% increase in the food fish supply in 1993. It has already been argued that this increase would have to be supplied mainly from aquaculture, which seems possible from a purely biotechnical vantage point. But whether it can occur will depend on socioeconomic if not political factors.

Health concerns will also play a significant role in the demand for fish. Many seafood products are low in fat and calories and high in protein and nutrients, and are considered good natural sources of omega-3 fatty acids, which have been linked with reduced rates of heart attacks, proper eye and brain development in infants, prevention and treatment of malaria, and maintenance of skin and tissue integrity in older people. In considering the health consciousness of consumers (especially in the West), many people will shift from red meat to fish; thus the projected increase in demand for fish is a conservative estimate.

Since the capture fisheries appear unable to keep pace with the growing demand for fish, aquaculture is regarded as an alternative source of fish supply. This is true especially in many developing countries where malnutrition, food insecurity, resource depletion, and environmental degradation are major concerns. However, conversion of cropland (such as rice fields) and diversion of irrigation waters for aquaculture may reduce the supply of essential foods (such as rice).

Foreign Exchange Earnings. Aquacultural products contribute a significant share of foreign exchange earnings in many countries (Shang, 1992b). The rapid expansion of shrimp and salmon culture are primarily export-oriented. However, the culture of exportable species, such as shrimps, usually requires the importation of hatchery equipment, feed, aerators, and technical expertise in many development countries. The net contribution of aquaculture to foreign-exchange earnings in those countries should be the difference between export value and import cost.

When aquacultural development is driven by foreign-exchange earnings, the decrease in fish products for domestic consumption may cause local hardships, especially in poor rural communities (such as the Philippines and Indonesia). In addition, the intensive large-scale culture of exportable species such as shrimps is often invested in by large outside investors such as multinational corporations. Large-scale intensive operations often create conflicts with small-scale farmers in competing for both inputs and outputs, resulting in uneven distribution of income and unemployment for unskilled labors. All of these have to be taken into consideration in government planning for aquacultural development.

Employment. Aquaculture generates many employment opportunities. Many of the people engaged in aquaculture are part-time laborers. Therefore, aquaculture is a valuable source of supplementary employment and income for rural people, especially for part-time fishermen-farmers and women or children. In addition, aquaculture generates spin-off employment in supporting industries such as feed, hatchery (or fry catching), processing, marketing, and so on. However, conversion of agricultural land and mangroves to commercial fishponds can displace peasants and landless laborers who lacked the basic skills to work on fish farms.

Recreation. People usually spend more time for leisure and recreation when personal income increases. In the Western countries, recreational fishery is a booming industry. For example, more than 26% of the U.S. population is engaged in recreational fishing, and the number is increasing steadily (Sandifer, 1988). Direct expenditures in the U.S. recreational fishing totaled $27 billion in 1985, nearly tripling the 1980 expenditure. The United States stocks over 844 million juvenile fish per year to support the recreational fisheries. These fish are produced in many public and private hatcheries. The recreational fishery creates employment and income opportunities.

Fishery Resources. In the face of the problems of overfishing due to capture fisheries, population pressures on land and coastal areas, and increasing demands for fish, a culture-based fishery is gradually regarded as an important alternative to increase the supply of fish. Culture-based fisheries involve producing hatchery-produced seed; nursing the intermediary stage; releasing juveniles into the ocean, lakes, or reservoirs; protecting them if possible; and then harvesting them as adults. This approach takes advantage of the open water itself, using it like large areas of aquatic pasture. Species ideal for sea ranching, for example, are migratory species such as salmon and relatively nonmobile species such as shrimps, lobsters, abalones, clams, and scallops. Japan is the most successful country in sea ranching. It releases about 2.5 billion fry of kuruma shrimp, scallops, abalones, clams, sea breams, and so forth (Liao, 1988) and about 2 billion salmon fry annually (Anonymous, 1990). Good results obtained in Japan indicate that sea ranching has a good future provided that jurisdiction over or ownership of the released stock can be established. In addition, sea ranching is intended to reduce the pressure on overfished stocks.

Aquarium Fish Trade. The aquarium fish trade is a multibillion-dollar industry in the world. The estimated retail value in 1992 was $3 billion (Bassleer, 1995). The major markets for aquarium fish are the United States, Europe, and Japan. Aquariums are found in about 8% of the estimated 86 million homes in the United States. In terms of popularity, the aquarium hobby is second only to photography (Winfree, 1989). In the United Kingdom, approximately 14% of the estimated 21 million British homes keep ornamental fish.

About 98% of the aquarium fish trade is in tropical fish with the remaining 2% in cold-water species. Ninety percent of the fish traded are freshwater, 10% are marine, and 0.1% are from brackish water. The available statistics cannot separate aquaculture from wild havesting, but the average unit price of aquarium fish is about 10 times of that of food fish (Bassleer, 1995).

The increasing demand for ornamental fish throughout the world has resulted in overexploitation of many natural stocks in the tropical countries. Recent technological advances have improved the capacity to induce spawning and rearing of aquarium fish under controlled conditions. Aquaculture will contribute significantly to the growth of the aquarium fish industry and remove pressure from wild population of many ornamental species. The industry has potential to provide an income for many people and to earn foreign exchange for many developing countries.

Industrial Purposes. The culture of seaweeds and pearls is mainly for industrial uses, which provide employment and income opportunities and earn foreign exchange.

Formulation of Development Plans (Regional and National). With aquacultural development objectives in mind, decision makers should make a comprehensive and critical review of the aquaculture sector, including resources availability, market potential, and supporting services. The resources needed (such as land, water, fry, feed, labor, electricity, and so forth) for the proposed development should be available at reasonable costs. Once the development objectives are identified and resource availability assessed, the development plan can be formulated and the proper species and systems for development can be decided on.

Species Selection. The species selected for culture depends on the purpose of aquacultural development in addition to the bioeconomic criteria mentioned earlier. In developing countries, staple species, such as carps, tilapias, mullets, and milkfish, are good candidates for domestic consumption, while high-valued omnivorous and carnivorous species, such as shrimps, salmons, and eels, are good candidates for export or for urban consumption. The tradeoffs between producing staple species and high-value species need to be weighed carefully. In developed countries, on the other hand, high-valued species with reasonable profit are usually selected for culture.

Increased production can be achieved by expanding production areas, improving existing culture techniques and practices, removing bottlenecks in supporting services, or a combination of these alternatives. The social benefits and costs of these alternatives should be assessed.

Culture Systems. The selection of culture systems (extensive, semi-intensive, and intensive) is usually determined by the availability and relative cost of inputs such as labor, capital, land, water, feed, and so forth. Extensive culture techniques are usually employed in places where there are an abundant supply of inexpensive labor and large areas of suitable land and water available at low cost, but where capital is relatively scarce. Such culture systems usually result in a low level of production per unit of water area, but the cost of production per unit of fish may also be low due to limited inputs. This kind of operation is especially appropriate if the objective of development is to produce cheap animal protein for local consumption and to increase employment opportunities in rural areas. The cultures of herbivorous species such as carps, tilapias, and milkfish in extensive and semi-intensive systems in many developing countries of Asia are good examples. In areas where land, water, and labor are relatively expensive compared with capital, intensive culture systems are usually employed, as in Japan, Israel, Taiwan, and the United States.

In addition to the relative costs of inputs, the environmental impacts of different culture intensities and systems should be an important consideration in selecting a sustainable farming practice. As mentioned earlier, extensive culture systems normally produce insignificant loading of nutrients or organic matter on the ecosystem, while intensive systems often create considerable pollution if not properly managed.

Polyculture and integrated farming are usually ecologically more sound than monoculture due to more efficient use of waste and energy on the farm. Farm waste can be recycled to produce multiple products, and water pollution can be reduced. These farming systems tend to be more favorable than monoculture from an environmental view point, especially in those areas with high prices of feed and energy and low labor costs. Also, the integrated farming systems are often more profitable (Bardach, 1986).

Cage and pen culture of fish is widely used for both freshwater and marine fish. Cage culture of sea breams, salmonids, seabass, snapper, grouper, tilapia, yellowtail, and carps and pen culture of milkfish and carps are practiced in various countries (Table 6.1). With improvements in structural design, materials, and farm operation, cage and pen fish culture is expected to expand in the future. However, cage and pen culture is also a source of organic pollution that contaminates culture areas. The overcrowding and resulting pollution of coastal cage culture in Japan have forced the Japanese fish farmers to venture farther offshore in an effort to raise healthy yellowtail and amberjack under semiwild conditions in unpolluted waters (see also Chapter 3).

Offshore marine aquaculture has attracted much attention recently. The main driving forces for offshore aquaculture are the environmental problems faced in inland and brackish-water aquaculture, the extensive open areas available in the outer coastal zone, the lower risk of disease transmission, the mobility of production units, and possibly lower production costs (Dahle, 1995). It is interesting to see more efforts being put into the development of offshore aquaculture, such as salmon culture in Norway and Canada, yellowtail culture in Japan, and seabass and sea bream culture in the Mediterranean. In offshore aquaculture, exclusive property rights to farming sites should be established. Otherwise, the investors would have little incentive to invest and to innovate (see Chapter 3).

Supporting Services. A sustainable aquacultural development project depends not only on increased production and income, but also on the existence of a potential market, an efficient marketing system, and other supporting systems such as hatcheries, feed mills, credit, research, training, and extension. All of these supporting services need to be well coordinated for sustainable development. If the capacity of

TABLE 6.1. Cage and Pen Fish Culture in Asia

Species	Cage/Pen	Country
Carps	Cage	China, Indonesia
Sea breams	Cage	Japan, South Korea, Taiwan
Groupers	Cage	Philippines, Hong Kong, Malaysia, Singapore, Thailand, China
Salmonids	Cage	Japan, China, South Korea
Seabass	Cage	Hong Kong, Indonesia, Thailand, Malaysia, Singapore
Snappers	Cage	Malaysia, Singapore, Thailand
Tilapia	Cage	Philippines, Singapore, Malaysia
Yellowtail	Cage	Japan, South Korea
Milkfish	Pen	Philippines
Carps	Pen	China, Thailand

these services is not adequate to handle the increased production, then expansion or improvement of these services should be included in the investment project. Shrimp and salmon farming, for example, are high-profit businesses and so attract continuous investment. However, the rapid increase in production of these species may soon depress their market prices. With increasing energy and feed costs, the farming of these species may encounter a cost-price squeeze, which is likely to slow down the speed of expansion. The industry needs to improve its production efficiency, to coordinate its production and marketing, and to diversify its markets and products in order to sustain growth. The success of the channel catfish industry in the United States is a good example of the coordination of production and marketing.

In many developing countries facing malnutrition and food insecurity, policy makers may feel that some deterioration in environmental quality is justifiable for economic development and that environmental conservation is a luxury practice they can ill afford (Dixon et al., 1988). Therefore, protecting the environment is not high on their priority list. An emphasis on planning for consumption now rather than for future generations is often preferred. But the tradeoff between consumption now and later should be carefully weighed. Environmental deterioration and misuse of scarce resources often result in real social losses in the long term and will undermine the basic requirements for sustainable development.

EXTERNALITIES AND METHODS OF VALUATION

The impacts of an aquacultural development project, both positive and negative, have to be identified and valued, and compared with alternative development opportunities in using limited resources. The negative impacts usually result from land- and water-use conflicts, effluent discharge, chemical residues, exotic species culture, and so forth (Table 6.2). For example, converting mangroves to fishponds is likely to result in a reduction in mangrove products and a loss of nursery grounds for

TABLE 6.2. Possible Socioeconomic and Environmental Impacts of Aquacultural Development

Activities	Possible Impacts
1. Conversion of mangroves for fishponds	Reduced mangrove products Reduced fisheries production Coastal erosion Unemployment created for unskilled labor Increased fish production in ponds
2. Conversion of cropland for fishponds	Reduced opportunity crop production Unemployment created for unskilled labor Shortage of essential food Increased fish production in ponds
3. Use ground and surface water	Reduced crop irrigation Land subsidence Saltwater intrusion into groundwater and agricultural fields Salinization of freshwater aquifers
4. Effluent discharge	Reduced downstream farm production Self-pollution Coastal or inland water pollution
5. Use chemicals, antibiotics, and hormones	Public health risks
6. New (exotic) species	Altered biodiversity Spread of diseases
7. Large-scale intensive culture	Conflicts with small-scale farmers Uneven income distribution Reduced employment opportunity for unskilled labor
8. Cage and pen culture	Reduced pressure on land and water Reduced fisheries yield in the same area due to pollution Conflicts with navigation, recreation, fishery, scenery, etc.
9. Demand for feed and fertilizer	Competition resulting in higher price of inputs to other activities Increased employment in these supporting industries
10. Sea farming	Preserved natural stocks Reduced pressure on land and water Increased marine fish production
11. Aquarium fish culture	Preserved natural stocks Increased export Employment effect
12. Increase aquacultural production (in general)	More fish and lower prices Increased employment in production, marketing, processing, etc. Increase in foreign exchange earnings Conflicts with other economic activities

fisheries (see also Chapter 4), coastal erosion, and unemployment for unskilled labor. Converting croplands to fishponds may reduce opportunity crop production (e.g., rice), result in a shortage of essential food, and create unemployment for unskilled labor. Using surface and groundwater for aquaculture may reduce crop production, create land subsidence, cause saltwater intrusion into groundwater and agricultural fields, and result in salinization of freshwater aquifers. Using public waters for farm waste disposal will reduce water quality and result in a loss of other water uses downstream. Using chemicals (pesticides, hormones, antibiotics, and so forth) may create health problems for human consumers. Introducing exotic species may reduce biodiversity through competition with indigenous species and inter-breeding, and increase the risk of transferring diseases. Cage and pen culture may create conflicts with other uses such as recreation, navigation, and scenery and may reduce fishery yields in the same areas due to pollution (see also Chapter 4). On the positive side, cage and pen culture, sea farming, and aquarium fish culture may release some pressure on the demand for land and water, preserve natural stocks, and increase employment in coastal areas.

In general, an increase in aquaculture production is likely to increase fish supply and reduce prices for consumers, improve living standards for some farmers, and increase foreign-exchange earnings. However, the expansion of aquaculture production will also result in competition between inputs (e.g., land, water, feed, and chemicals) with other economic activities; this in turn results in higher prices of inputs to other industries in addition to the other possible conflicts mentioned earlier (see also Chapter 8).

Methods of Valuation

After both the economic and environmental impacts of an aquacultural development project are identified, the most important task for the economic analysis is to monetize them. Social cost-benefit analysis is commonly used for economic viability studies. Both direct and indirect benefits and costs of the project have to be estimated. Several guidelines are useful in estimating the benefits and costs of a project (Dixon et al., 1988; Shang, 1990):

1. Changes in productivity caused by the project both on farm and off farm have to be evaluated. Changes off-farm include all economic and environmental externalities (both positive and negative). Physical changes in productivity are valued either using market price for inputs and outputs or using shadow price when distortions exit. Distortions often arise as a result of taxes, subsidies, exchange rates, mandated wages, and so on. An aquaculture project, for example, may disrupt downstream farm (fish or agricultural) activities. In economic evaluation, only the net change of the related activities should be used.
2. The benefits and costs should be estimated on a with-and-without-project basis. Only the additional or incremental benefits and costs due to implementation of the project should be considered.

3. A benefit forgone due to the implementation of the project is treated as a cost, while a cost avoided is a benefit.
4. Loss of earnings and medical costs caused by a project or the comparable savings that would accrue from preventing that damage are often used to evaluate human capital.
5. An opportunity cost approach is often used to estimate for the cost of using the resource for aquaculture, which is the cost of the forgone income from other uses of the resources (e.g., land, water, labor, and so on).
6. Approaches that consider replacement costs and prevention costs are useful in cost-benefit valuation. The costs of replacing or preventing the damages caused by the project can be estimated and incorporated into the analysis.
7. It is generally agreed that the discount rate used in the social cost-benefit analysis should be linked to the marginal productivity of capital. A low discount rate will avoid excessive resource use in the present, while a high discount rate will push toward present use resulting in quick exhaustion of the resource.

In addition to the social cost-benefit analysis, input-output analysis and linear programming can be used to evaluate the direct and indirect impact of aquaculture projects on the economy and the environment (Dixon et al., 1988; Cerezo and Clonts, 1994).

Recently, an aquacultural development decision support system has been applied to aquaculture (El-Gayar, 1995). This approach seeks to optimize allocation of resources and activity levels that would strike an acceptable balance among various developmental goals subject to the constraints of resources, market, and pollution. It would systematically aid decision makers to plan for aquacultural development in a region. The details of this approach can be found in Chapter 7.

Mitigation Measures

Most aquacultural development activities create some adverse impact on the environment. The environmental externalities or spillovers are important factors influencing the sustainability of aquacultural development projects. When an environmental externality exists, this implies that the private costs of a commercial farm are less than its social costs. Consequently, if the farm aims to maximize its private gain or profit, it does not take into account its adverse spillovers. Economic intervention in this case is often needed to bring the private costs of the farm into line with its social costs, thereby internalizing the environmental externality (Tisdell, 1995). All of the public and private mitigation alternatives (Figure 6.5) need to be evaluated economically, and the most cost-effective measures should be adopted.

Public Measures. The most popular policy forms recommended for reducing externalities are taxes, subsidies, and standards (Cerezo and Clonts, 1994). Taxes here refers to fees collected by the public agency and imposed on each unit of pollutant discharged into public waters. Taxes on the use of some inputs that can indirectly

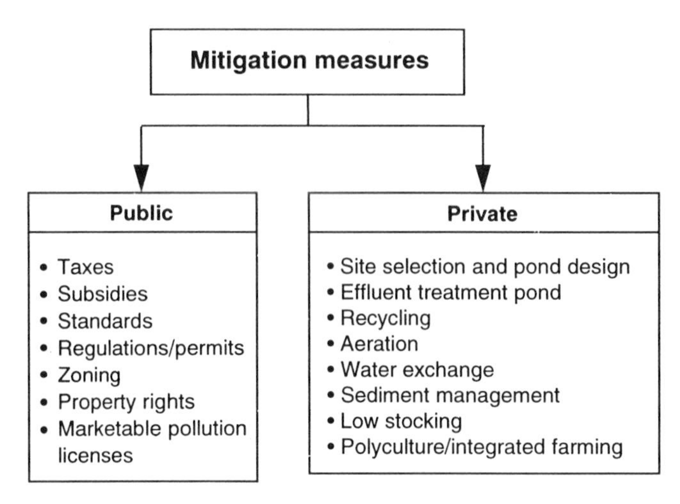

Figure 6.5. Measures that mitigate externalities.

have unfavorable externalities can also be levied. Subsidies are often used in the form of lower input cost for waste treatment or pollution reduction, or tariff exemption on imported pollution control equipments. Standards establish maximum acceptable levels of waste discharged. These measures tend to induce fish farmers to find cost-effective methods to meet environmental constraints for sustainable development. In many cases, a combination of these measures such as tax and standards, may be most effective. A tax would be levied if a fish producer's discharge is above the specified limits. A profit-maximizing fish farmer usually reduces waste discharge until the marginal cost of waste reduction equals the cost (tax penalty) of not treating the waste.

Regulations and permits are the other public means of preventing adverse impacts. Permits are often required to use land and water for aquaculture, to use particular chemicals on the farm, to discharge effluents into public waters, to import exotic species, and to convert mangroves into fishponds. Regulations on stocking rates, the use of some inputs, the use of particular farming technologies, water-quality parameters in effluents, the ingredients of feed, and so forth, are also practiced.

Zoning and allocation of sites for aquaculture are often practiced in order to avoid conflicts of land and water use and to prevent or reduce pollution. Establishment of private property rights is also recommended as a way to reduce spillovers. Property rights provide incentives for investment in sustainable development. Legal rights to aquaculture sites can be sold (by auction) or leased to private operators. However, exclusive private rights to water colume are difficult to establish when large water bodies are shared in common. In Japan, legal rights to coastal aquaculture sites belong to fish farmers' (or fishermen's) associations. The associations then allocate the right to their members. This approach seems to be working well in solving the legal-rights problems in commonly shared large water bodies.

Marketable licenses and permits for environmental use have been promoted in recent years as a means of dealing with environmental spillovers. However, this measure may be of limited value for aquaculture because many of externalities involved are location-specific rather than geographically uniform (Tisdell, 1995).

It is important to mention that most of these measures involve monetary and enforcement costs and may also be difficult to apply. The benefits generated by these measures may not be justified by their costs. These factors have to be considered in economic decision making.

Private Measures. At a farm level, good farm management practices can reduce many of the adverse impacts. These practices include using good site selection and design; adding effluent treatment ponds; using water recycling and aeration; keeping stocking rates low; using biofilters, dilution, and sediment management; maintaining adequate distance between groups of cages; using polyculture or integrated aquaculture-agriculture farming rather than monoculture among others. Generally, these measures can be carried out only if they are technically and economically feasible, and if reasonable loans and adequate extension services are available.

The cost-effectiveness analysis is often used to select the best mitigation measure after considering a number of alternatives and their costs. The major difference between this approach and the cost-benefit analysis is that no attempt is made to monetize benefits. Rather, the goal is to identify the least-cost alternatives that will achieve the selected goal or target.

Some developments have occurred in environmental economics and in policy discussions that could result in increasing use of cost-effectiveness methods in the future for assessment of aquacultural developments. These are that (1) it may be necessary to impose strong conditions on the conservation of natural environmental resource stocks to achieve sustainable economic development; (2) because of uncertainty, a precautionary attitude toward the conservation of natural resources is needed; and (3) the increasing prevalence of ecocentric attitudes toward nature conservation relative to anthropocentric ones may constrain traditional cost-benefit analysis.

DISCUSSION

The above aspects imply that while environmental externalities should be taken into account in project assessment, this is not sufficient. Advocates of strong conditions for economic sustainability argue that the natural environmental resource stocks of the world have been reduced to critically low levels and that further reduction will reduce the incomes of future generations (Peare, 1993). Thus policies must be devised to conserve existing natural environmental resource stocks. Such policies need not, however, result in the absence of economic development, including the absence of aquacultural development. Project development can proceed, for instance, by the use of offset policies (Tisdell, 1993). If, for example, an aquacultural development would lead to the loss of a natural wetland, the development may be permitted if a suitable compensatory artificial wetland is created. While environmental externalities should be taken into account in such policy deliberations, social cost-benefit analysis does not determine the scale of operations because a macroenvironmental constraint is imposed.

The precautionary motive can also result in restrictions on the use of social cost-benefit analysis. For example, safe minimum environmental standards may be a constraint. Furthermore, social cost-benefit analysis, like all existing major methods of economic evaluation, is anthropocentric. Expression of ecocentric values may restrict its application. The decision, for instance, in Denmark to limit the maximum nitrogen content of manufactured fish food (New, 1995) because of the impacts of wastes from aquaculture on natural ecosystems seems in part to be an expression of ecocentric values—a strong desire to sustain natural ecosystems.

Of course, not all economists are convinced that strong sustainability conditions are needed to ensure sustainable economic development. Some believe that adequate accumulation of manmade capital is all that is needed, and they remain optimistic about the ability of the traditional formula for economic growth to deliver sustainable economic development. The adoption of new knowledge or environmentally beneficial technology, such as the replacement of monoculture by appropriate polycultures, can contribute to sustainable development. These economists propose weak conditions for attaining economic sustainability. Nevertheless, advocates of this approach also recommend that environmental externalities be taken into account in project evaluation. In addition, the intertemporal impacts of projects are not ignored, although there is argument of what constitutes the appropriate rate of time discounting or whether any discounting should be applied (see, e.g., Kula 1992). Yet it is clear that both approaches (weak and strong) have policy implications for aquaculture and both require taking into account environmental impacts and externalities when assessing the sustainability of aquaculture projects.

The World Commission on Environment and Development (WCED) (1987) provided an optimistic view of the ability of progress in aquaculture to contribute to global sustainable development. While advances in aquaculture have an important role to play in sustainable development, this report gives insufficient attention to the requirements that must be met for sustainability of aquaculture projects (Tisdell, 1991). Such projects must be assessed not only in terms of their economics, but also in terms of their social and biophysical impacts to ensure that they contribute to sustainable development.

In summary, economics plays an important role in sustainable aquacultural development. On the farm level, an optimal management strategy is needed to produce the best sustainable net income without degrading the environment. The externalities and socioeconomic feasibilities of aquacultural development projects need to be assessed to guide the sustainable development on society at large. Various public and private measures can be used to reduce the negative externalities, but their economic viability needs to be assessed, and the most cost-effective ones should be adopted.

Sustainable aquaculture requires adequate consideration of the interactions among socioeconomic, biotechnical, and ecological factors. It also requires a balanced development in production, processing, marketing, and policy matters. Careful management of aquaculture and of the economic sectors interacting with it is needed if a socially appropriate degree of sustainability in production and economic welfare is to be achieved. It is important to realize that natural resources will have to be shared with all potential users in a way that will benefit the society while not harming the ecosystem.

REFERENCES

Anonymous. 1990. Salmon culture. Yamaha Fish. J. (Shizuoka-ken, Japan), 32: 1–8.

Bardach, J. E. 1986. Constraints to polyculture. Aquacult. Eng. 5: 287–300.

Bassleer, G. 1995. The international trade in ornamental fish. pp. 241–243, In: Aquaculture towards the 21st century. Proceedings of Infofish-Aquatech '94 International Conference on Aquaculture. August 29–31, 1994, Colombo, Sri Lanka.

Cerezo, G. A., and H. A. Clonts. 1994. Economic analysis of effluent control from catfish ponds. Alabama Agricultural Experiment Station, Auburn University. 38 p.

Cheng, H. T., and O. Capps, Jr. 1988. Demand analysis of fresh and frozen finfish and shellfish in the United States. Am. J. Agric. Econ., pp. 533–542.

Csavas, I. 1994. Recommendations for sustainable aquaculture. Presented at the SEADEC/AQD Seminar-Workshop of Aquaculture Development in Southeast Asia, Iloilo City, Philippines.

Dahle, L. A. 1995. Offshore fish farming-recent development. pp. 169–184, In: Kpp. Nambiar Tarlochan Singh (Ed.). Aquaculture towards the 21st century. Proceedings of Infofish-Aquatech '94 International Conference on Aquaculture, August 29–31, 1994, Colombo, Sri Lanka.

Dixon, J. A. Carpenter, R. A. Fallon, L. A. Sherman, P. B., and Manopimoke, S. 1988. Economic analysis of environmental impacts of development projects. Earthscan Publications, London. 134 p.

El-Gayar, O. F. A. 1995. An aquaculture development decision support system. Dissertation, University of Hawaii, Honolulu. 208 p.

FAO. 1995a. Aquaculture production statistics, 1984–1993. Food and Agriculture Organization of the United Nations, Rome. 186 p.

FAO. 1995b. The state of world fisheries and aquaculture. Food and Agriculture Organization of the United Nations, Rome. 57 p.

Kula, E. 1992. Economics of natural resources and the environment. Chapman & Hall, London. 287 p.

Liao, I. C. 1988. East meet west: an eastern perspective of aquaculture. J. World Aquacult. Soc., 19(2): 73–84.

New, M. 1995. Aquafeeds for sustainable development. Paper presented at Sustainable Aquaculture '95: The Pacific Congress on Marine Science and Technology, June 11–14, 1995, Honolulu, Hawaii.

Peare, D. 1993. Blueprint 3: measuring sustainable development. Earthscan Publications, London. 224 p.

Sandifer, P. A. 1988. Aquaculture in the west, a perspective. J. World Aquacult. Soc., 19(2): 73–84.

Shang, Y. C. 1981. Aquaculture economics: basic concepts and methods of analysis. Westview Press, Boulder, Colo. 153 p.

Shang, Y. C. 1990. Aquaculture economic analysis: an introduction. World Aquaculture Society, Baton Rouge, 221 p.

Shang, Y. C. 1992a. Prospects of aquaculture development in Asia. Paper presented at the Third Asian Fisheries Forum, October 26–30, 1992, Singapore.

Shang, Y. C. 1992b. The role of aquaculture in the world fisheries. Paper presented at the World Fisheries Congress, May 3–8, 1992, Athens, Greece. Oxford and IBH Publishing Co., New Delhi, India.

Tisdell, C. A. 1991. Development of aquaculture and the environment: coastal conflicts and giant clam farming as a case study. Int. J. Environ. Stud., 39: 35–44.

Tisdell, C. A. 1993. Environmental economics. Edward Elgar, Aldershot, U.K. 259 p.

Tisdell, C. A. 1995. The environment and sustainable aquaculture. pp. 384–393, In: Proceedings of Sustainable Aquaculture '95. The Pacific Congress on Marine Science and Technology, June 11–14, 1995, Honolulu, Hawaii.

WCED. (World Commission on Environment and Development). 1987. Our common future. Oxford University Press, New York. 400 p.

Winfree, R. A. 1989. Tropical fish: their production and marketing in the United States. World Aquacult. 20(3): 24–30.

Yamaha Fishery Journal. 1990. Salmon culture. Shizuoka-ken, Japan. No. 32, pp. 1–8.

7 The Role of Modeling in the Managing and Planning of Sustainable Aquaculture

PINGSUN LEUNG AND OMAR F. EL-GAYAR

INTRODUCTION

A model is a simplified representation of reality. By *simplified* we mean that a model contains only those attributes that are relevant to a particular problem. Models are useful tools for representing and experimenting with complex reality. *Modeling* thus refers to the act of constructing, validating, and using models.

Cuenco (1989) provides the following answers to the question, Why is there a need to model aquacultural systems, one of many hybrid natural-social-technical systems?

1. Modeling serves as a powerful tool for the formulation, examination, and improvement of hypotheses and theories.
2. Models can make intelligent predictions about the consequences of various management strategies on the system.
3. Modeling provides a working tool to conduct numerous "what if" experiments quickly and to evaluate the consequences of various hypotheses or management strategies of large and complex aquacultural systems, which are seldom possible in their natural environment.
4. Models serve as mechanisms to identify what is not known by organizing what is known within a framework.
5. Models facilitate the evaluation of complex interactions of aquacultural systems.
6. Modeling accelerates the use of more quantitative and precise methods in aquacultural research.
7. Models put together knowledge from theoretical, laboratory, and field studies into a consistent whole so as to identify areas where knowledge is lacking, sparse, or inconsistent.

Sustainable Aquaculture, Edited by John E. Bardach
ISBN 0-471-14829-6

To summarize, a model is used because of its economy, availability, and ease of information handling. It may cost less to derive knowledge from a model than from the real-world counterpart such as a large, complex aquacultural system. A model may represent an aquacultural system that does not exist or cannot be easily manipulated. In addition, a model may provide a convenient medium to collect and/or transmit information. Models of aquacutural systems are largely extensions of models of agricultural systems, with special attention to the handling of the growing medium.

The current drive toward sustainable aquaculture is expected to emphasize further the need and importance of modeling aquacultural systems. Of particular importance are models that seek to improve land, water, energy, and feed use as well as to decrease the pollution potential of aquaculture. As Chamberlain and Rosenthal (1995) indicate, sustainability as a concept refers to management practices that will not degrade the environment. Modeling is an alternative for the development and evaluation of such management practices.

The next section provides an overview of aquacultural modeling with particular emphasis on management and planning applications of bioeconomic models at the farm as well as the regional level. In the section "Modeling Sustainability" we discuss the need and importance of tackling sustainability as a multicriteria concept as opposed to the traditional single-criterion models. A decision support system for aquacultural development is presented. The use of this system in analyzing the tradeoffs among economic, social, and environmental goals is also demonstrated through a case study in Northern Egypt.

AN OVERVIEW OF AQUACULTURE MODELING

While models developed for aquaculture can be classified in several ways based on the specific model characteristic chosen for classification, it is instructive to broadly categorize aquacultural models as biological, economic, biophysical, and bioeconomic, as this scheme emphasizes and clarifies the underlying approach on which the model is built. Each of these model types is described below with a brief introduction to the different subtypes in each broad category.

The overview also includes a survey of bioeconomic models as well as a brief description of a sample of such models. Particular emphasis is given to bioeconomic models, as these already include the biological and physical elements for the economic evaluation of aquacultural systems. In the pursuit of sustainable aquaculture a primary goal is to develop aquacultural systems and management practices that are biologically and physically feasible, economically viable, and environment-friendly. Bioeconomic models represent a merger of the other types of models and thereby tend to be most useful for development and evaluation of sustainable aquacultural systems and management practices at both operational and strategic levels.

Biological Models

Biological models are primarily concerned with modeling the biological processes involved in aquaculture-related activities. Such models are generally difficult to construct because of the complexity of the biological organism and its interactions with the environment. In many cases, particularly in pond culture, the aquacultural manager has very little knowledge of how the production in the pond is progressing. Unlike traditional agriculture or terrestrial livestock production, the aquacultural producer cannot directly visualize the growth of the "crop" and therefore must rely on indirect and subjective measures of production on which to base management decisions. Hatch and Kinnucan (1993) point out three crucial factors in aquaculture production that are not essential for other animal production systems—namely, pond ecology, monitoring of animals, and feed utilization—and these also require additional modeling sophistication.

In effect, the aquacultural pond can be aptly described as a black box. Biological models try to unveil the contents of this black box to a certain degree, in a sense to make it somewhat "translucent." It is important since the biological model component generally forms the hub of the total bioeconomic model in that it defines the requirements of the other components. Piedrahita (1988) has classified computer models of aquatic ecosystems as either empirical or mechanistic. Bernard (1983) has categorized mathematical models of pond aquaculture into three major types: empirical, stock, and mechanistic. Cuenco (1989) has classified the biological models further into growth models, pond ecosystem models, production models, survival models, dissolved-oxygen models, ammonia models, water-temperature models, and salinity models. For the purpose of classifying the biological models used in this survey, we have used Bernard's classification. However, as both Piedrahita and Bernard have indicated, actual models combine attributes of all three types and the dividing lines among the three types of models are not clear-cut.

Empirical models set out principally to describe situations in which the ecosystem or culture environment is treated as a black box with only inputs and outputs. Processes that take place inside the system are ignored. To a certain extent, empirical models are similar to production function analysis in economics, where outputs are treated as functions of inputs without detailed depiction of the underlying biological relationships.

Stock models generally separate the stock or population dynamics into growth, mortality, reproduction, and recruitment of stock numbers. Stock models lie somewhere in between empirical and mechanistic models.

Mechanistic models attempt to give an understanding of the biological and environmental processes of the ecosystem or culture environment on a finer scale than do stock models. Mechanistic models are complex syntheses of what is known of the ecosystem and hence are very difficult to build and use. They are sometimes referred to as being "internally descriptive" or "theoretical" models. Mechanistic models of aquatic ecosystems are based primarily on mass balances, while fish growth models use energy balances (Piedrahita, 1988). Figure 7.1 shows the three types of models.

It is always possible to find an empirical model that gives a better fit to a given set of data than a stock or mechanistic model does. This arises because the empirical model has fewer constraints, whereas a stock or mechanistic model can be constrained by its assumptions, even when it contains more adjustable parameters. As concluded by Bernard (1983), empirical and, to a certain extent, stock models can be

(a) Empirical Model

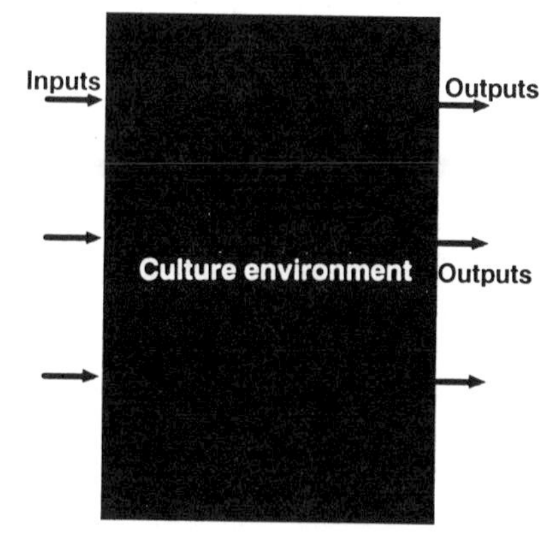

(b) Stock Model

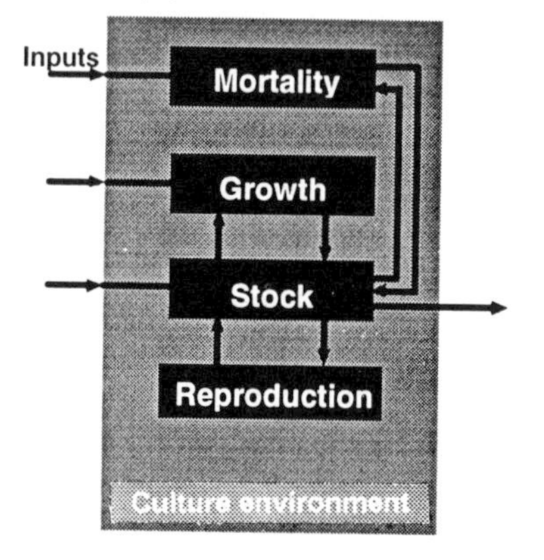

(c) Mechanistic Model

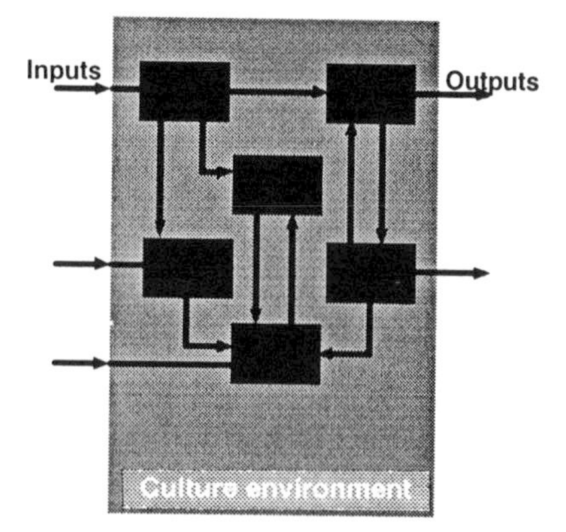

Figure 7.1. Schematics of the three types of biological models.

adapted for management. Mechanistic models, on the other hand, are not well suited for management purposes as they are rather difficult to build and use. However, they can provide insight into the precision of empirical or stock models, and can help direct future research efforts to improve that precision.

Economic Models

The choice of appropriate economic models for economic evaluation of aquacultural systems depends on the problems to be solved and the amount of compatibility needed with the biological models. Following Allen et al. (1984), economic models can be generally classified as optimizing or nonoptimizing. *Optimizing models* refer to techniques that seek the best value of an objective function expressed in terms of a set of control variables in an aquacultural system. Differential calculus is used when there is no constraint, and Lagrangian functions are used when the constraints are equalities. In the event that analytical solutions cannot be derived, iterative numerical (search) methods are used. In this survey, we refer to these methods as *classical optimization* techniques. The Faustmann model, originally developed for optimal forest rotation (Faustmann, 1849), can also be applied to solving the problem of optimal harvesting age or rotation period for a single growing population of aquatic organisms. In this chapter, the Faustmann model is lumped together with classical optimization.

When the constraints are inequalities, mathematical programming techniques are used. By far the most commonly used mathematical programming technique is linear programming. *Linear programming* is the process of determining the values of variables that optimize a linear objective function while satisfying a set of linear constraints. Linear programming is widely used in farm planning for choosing a profit-maximizing combination of production alternatives that is feasible with respect to a set of fixed farm constraints. *Risk programming* models are standard methodology for extending the linear programming framework to include farm risk. Commonly used models include quadratic programming, which minimizes the return variance of the farm portfolio, and MOTAD (minimization of total absolute deviations) (Hazell and Norton 1986). Tauer's (1983) *target MOTAD* model extends the MOTAD model to include a minimum return target. Including this type of safety-first feature is most appropriate when the risk of catastrophe is large, either because of an inherently risky environment or because the farmer is poor and has minimal reserves to fall back on in a bad year. *Integer and mixed-integer programming* models are called on when decisions involve indivisible objects such as tractors, buildings, ponds, and tanks.

When it is desirable to optimize over time, multiperiod linear programming or *dynamic programming* can be used. They are particularly suitable for aquaculture production since decisions are usually time-dependent. As most of the biological as well as economic relations in the real world are highly nonlinear, *nonlinear programming* provides a more realistic alternative to model aquaculture production systems. It is becoming quite popular as more reliable nonlinear programming software becomes available. *Optimal control theory* extends the classical techniques to cover both nonlinear and linear optimization problems and inequality constraints. It provides a powerful analytical tool for handling dynamic (time-dependent) systems and possesses valuable economic interpretations. The dynamic programming model is often viewed as the discrete time version of the optimal control model. As Cacho (1993) has put it, "dynamic programming is better suited to study more realistic stochastic problems." In other words, optimal control theory provides the theoretical basis while dynamic programming provides the tool for empirical analysis.

Optimization techniques require tractable functional forms, and the realism of the model is often sacrificed. *Simulation* can be used when it is not possible to express the interrelation in a convenient mathematical form, either because the system is too complex or because responses are subject to random variations. Simulation merely describes the output behavior of different combinations of inputs, control variables, and parameter values. It is generally nonoptimizing and usually requires a large amount of computer time. For the purpose of this survey, simulation is classified into budget simulation and process simulation. *Budget simulation* usually uses very simple biological models but includes extensive financial analysis capabilities such as enterprise budgeting and cash-flow or capital budgeting. *Process simulation*, however, provides rather detailed biological process simulation using either general-purpose computer languages or special-purpose simulators. There are many economic budget analyses of aquaculture, but only those with a recognizable biophysical model component are included in this chapter.

In summary, optimization models require tractable functions but yield the best solution, while simulation can use more realism but may not find the best solution and may be very expensive in terms of computer time. A hybrid of these two classes of models can be helpful in some cases. Figure 7.2 shows the basic types of economic models.

Biophysical Models

Biophysical models are primarily concerned with modeling the biological and physical aspects of an aquacultural environment without taking the economic issues into consideration. Numerous biophysical models have been developed for aquaculture (Allen et al., 1984; Cuenco, 1989; Piedrahita, 1988). While biophysical models can be of value for research as well as engineering purposes, the fact that they lack the economic component renders such models of limited use in the development of sustainable aquacultural systems and management practices.

Bioeconomic Models

Allen et al. (1984) have defined the term *bioeconomics* as the use of mathematical models to relate the biological performance of a production system to its economic and technical constraints. They emphasize the systematic integration of the three functionally important and related areas of aquaculture production systems: biological performance, physical systems, and economic considerations. A bioeconomic model can thus be viewed as a mathematical description of an aquaculture enterprise

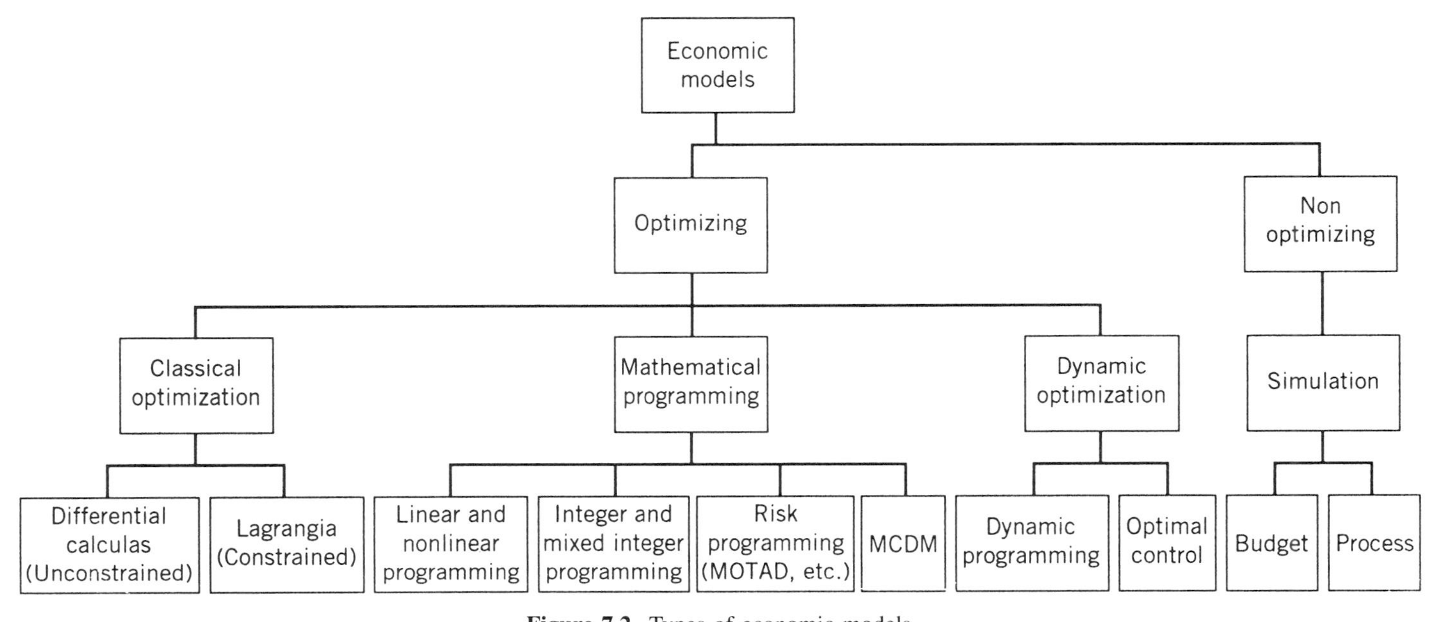

Figure 7.2. Types of economic models.

by interrelating its biological, physical, and economic elements for a holistic and systematic analysis.

Bioeconomic modeling provides an alternative method for representing the production process as compared with conventional production function analysis. It allows evaluations of a wider range of environmental conditions than would normally be possible with purely economic models since the biotechnical relationships can be more clearly defined. Bioeconomic modeling thus offers the greatest potential for the development and planning of sustainable aquaculture.

It is also worth noting that bioeconomic models for aquacultural systems are few in comparison with agricultural systems and that their existence is less than a quarter century old. The seminal work by Allen et al. (1984) provided a comprehensive survey of aquacultural bioeconomic models developed in the decade before 1984. On the other hand, Leung (1993) has provided a comprehensive survey of the models developed during the last decade. Table 7.1 presents a list of bioeconomic models developed during the period 1974 to 1994. Models in the list are broadly grouped by application. The following subsections present brief descriptions of a sample of bioeconomic model applications at the farm and regional levels.

Applications of Bioeconomic Models at the Farm Level. Bioeconomic models have been applied mostly at the farm level. In Table 7.1, farm-level applications are broadly categorized as management applications, feasibility analysis applications, design applications, and research applications.

Management applications are primarily concerned with the day-to-day management of aquaculture operations, such as determining optimal stocking and harvesting schedules, determining optimal feed and fertilizer applications, and examining the effect of different management strategies. As an example, Bjorndal (1988) has developed a bioeconomic model to analyze the optimal harvest time for farmed fish. The biological model is built on the basic fact that over time, stocked fish will grow and some will die. Growth is expressed as a function of time (age), number of fish (density), and feed quantity. Mortality rate is assumed to be constant. Thus, the biological model represents the changes of the stock of fish over time as a result of growth and natural mortality. Given the biological constraints, the economic model applies optimal control theory to find the harvesting time that maximizes the present value of profits. Variable costs considered in the model are feed, harvest, and insurance costs. The results indicate that one should harvest the fish later when there are harvesting costs than when there are no such costs. Also, the difference in growth rates can cause selective harvesting to be optimal. Both results have a strong resemblance to the optimal exploitation of renewable resources, particularly in forestry.

Feasibility and economic analysis applications of bioeconomic models are primarily concerned with evaluating the financial and economic viability of aquacultural systems and management practices by taking into consideration the biological and physical constraints imposed. Of particular relevance to making aquaculture sustainable is the use of a bioeconomic model by Alsagoff et al. (1992) to analyze the economics of introducing integrated aquacultural farming (IAF) to small-scale

TABLE 7.1. List of Bioeconomic Models Developed during the Period 1974 to 1994

Model	Description	Biological Models	Economic Models
	Management Applications		
Sparre (1976)	Optimal stocking, feeding and harvesting decisions	Mechanistic	Dynamic programming—Markov chain[a]
Talpaz and Tsur (1982)	Optimal harvesting strategy	Stock	Optimal control
Azizan (1983)	Optimal harvesting	Stock	Simulation
Pardy et al. (1983)	Evaluating alternative stocking strategies	Stock	Simulation
Tsur and Hochman (1986)	Optimal management (harvesting policy) of algal biomass	Stock	Optimal control
Karp et al. (1986)	Optimal stocking and harvesting	Stock	Dynamic programming
Bjorndal (1988)	Optimal harvesting of farmed fish	Stock	Optimal control; Faustmann model[a]
Leung and Shang (1989)	Optimal stocking and harvesting strategies	Stock	Dynamic programming—Markov chain[a]
Hochman et al. (1990)	Optimal harvest schedule	Stock	Dynamic programming—Markov chain[a]
Springborn et al. (1992)	Optimal harvest time	Stock	Simulation
LaFranchi (1992)	Optimal species selection and harvest schedule	Stock	Nonlinear programming
Tisdell et al. (1993)	Optimal length of harvest cycle	Stock	Classical optimization
Summerfelt et al. (1993)	Optimal stocking and harvesting	Stock	Simulation
Leung et al. (1994)	Optimal harvest (rotation) age	Stock	Faustmann model
Clark (1994)	Optimal harvest for integrated farming systems	Mechanistic	Nonlinear programming
Hean (1994)	Optimal management strategy	Stock	Optimal control
Johnson (1974)	Schedule of release dates and choice of stocks in hatchery	Stock	Linear programming
Botsford et al. (1974)	Optimal temperature for lobster culture	Stock	Optimal control
Emanuel and Mulholland (1975)	Optimal schedule of fertilizer application	Mechanistic	Optimal control
Bala and Satter (1989)	Optimal operation and management	Stock	System dynamics and linear programming

(continued)

TABLE 7.1. (continued)

Model	Description	Biological Models	Economic Models
	Management Applications (cont.)		
Sadeh et al. (1989)	Optimal management	Stock	Optimal control
Botsford (1977); Botsford et al. (1977)	Choice of heat sources	Mechanistic	Optimal control; simulation
Alsagoff et al. (1990)	Synchronizing broiler and fish growth cycles for organic fertilization of fishponds	Empirical	Mixed-integer programming model
Shaftel and Wilson (1990)	Production scheduling and strategic planning w.r.t. technological changes	Empirical	Mixed-integer linear programming model
Cacho et al. (1990); Cacho et al. (1991)	Cost-effective feeding regimes for pond-reared fish	Mechanistic; bioenergetic	Optimal control
Sadeh (1986)	Value of information in production process	Stock	Optimal control; dynamic programming
Hatch et al. (1987)	Evaluating production choices	Empirical	Target MOTAD
Hatch and Atwood (1988)	Assessing risk-income tradeoffs for alternative production activities	Empirical	Target MOTAD
Engle and Hatch (1988)	Aeration strategy	Empirical	Target MOTAD
Hatch et al. (1989)	Assessing risk-income tradeoffs for alternative production activities	Empirical	Discrete stochastic programming model—chance constraints
Engle and Pounds (1993)	Optimal production management strategies—single batch vs. multiple batch	Empirical	Multiperiod target MOTAD
Logan and Shigekawa (1986)	Size and product mix of commercial sturgeon production	Stock	Simulation
Wu (1989)	Evaluating candidate species for mariculture	Stock	Budgeting
Logan and Johnston (1993)	Optimum replacement of broodstock for rainbow trout	Stock	Simulation—replacement analysis
El-Gayar and Leung (1994)	Choices of species	Empirical	MOTAD

	Feasibility Analysis Applications		
Gates et al. (1980a, 1980b)	Optimal fish culture decisions in a water reuse system, and financial feasibility analysis	Stock—Markov chain[a]	Dynamic linear programming
Lipschultz and Krantz (1980)	Combination of purchases and culture activities that minimized total culture cost	Empirical	Linear programming
Adams et al. (1980a, 1980b)	Financial feasibility of shrimp grow-out operation	Stock	Budget simulation
Griffin et al. (1983)	Generalized budget simulation system	Stock	Budget simulation
Van Hemelryck (as described in Allen et al. 1984)	Cost of producing the Japanese clam	Stock	Simulation
Hanson et al. (1985)	Financial performance of shrimp farms	Stock	Budget simulation (MARSIM)
Leung and Rowland (1989)	Financial evaluation of alternative design and operation	Empirical	Budgeting—electronic spreadsheet
Alsagoff et al. (1992)	Evaluating the economics of integrated agriculture-aquaculture mixed farming	Empirical	Linear programming
Gempesaw et al. (1993)	Economics of vertical integration in striped bass aquaculture	Empirical	Budget simulation (AQUASM)
Bacon et al. (1993)	Evaluating economic benefits of incorporating a small-scale trout enterprise with a grain and broiler farm	Empirical	Budgeting simulation (AQUAISM)

(*continued*)

TABLE 7.1. (continued)

Model	Description	Biological Models	Economic Models
	Design Applications		
Schuur et al. (1974); Botsford et al. (1975)	Optimal culture systems: stacked tray, raceway, silo	Mechanistic	Optimal control; simulation
Rauch et al (1975)	Design and operation of an aquaculture facility	Mechanistic	Optimal control
McNown and Seireg (1983)	Optimal aquacultural plant design	Stock	Dynamic programming
Tian (1993)	Optimal aquafarm structure and size	Empirical	Nonlinear programming
Callaghan (1975)	Evaluating indirect heating from electric generating station thermal effluent	Stock	Simulation (DYNAMO)
Syed (1985)	Comparison of different culture systems	Empirical	Mixed-integer programming model
	Research Applications		
Allen and Johnston (1976); Johnston and Botsford (1981)	Defining areas in which research success would have high potential rewards	Mechanistic	Simulation
Griffin et al. (1981)	Directing future research needs	Mechanistic	Simulation
Griffin et al. (1984)	Directing future research	Mechanistic	Simulation
	Regional-Level Applications		
Sylvia and Anderson (1993)	Optimizing public and private net-pen salmon aquacultural strategies	Empirical	Dynamic multilevel programming model

[a]Markov chain is useful in studying the evolution of systems over time where the outcome of the system cannot be determined with certainty. It is often used to model the probabilistic growth process of aquaculture species and is commonly used in conjunction with dynamic programming.

farmers in Central Perak, Malaysia. The importance of IAF is that it integrates traditional farming with animal husbandry and fish polyculture thereby minimizing waste discharge. The economic model is formulated as a linear programming (LP) problem where the objective is to maximize annual farm income subject to resource and biological constraints. Resource constraints include land, labor, and capital, while biological constraints include fish stocking densities and poultry housing capacity. The results indicate that annual farm income would increase threefold with the introduction of IAF.

Bioeconomic models are also utilized in the optimal design of aquacultural systems by taking the biological and physical constraints into consideration. For example, McNown and Seireg (1983) developed a computer program based on a bioeconomic model for the optimal design of a multistage recirculating aquacultural system. The biological component was a fish-growth model in which growth is a function of fish size, temperature, and feeding rate. The economic model utilized the dynamic programming approach. The objective was to minimize the total cost of growing fish by determining design parameters such as the number of pieces of equipment required and the number and size of fish produced, as well as operating parameters such as the culture temperature, feeding rates, and densities to be used for each stage. Particular emphasis was given to energy utilization by incorporating the effects of insulation, season, location, solar utilization, and changes in operational parameters or market conditions into the model. The results indicate that staging the grow-out process for perch and rainbow trout produces a cost reduction of at least 30%. It should be noted that although the model development was primarily driven by the increasing cost of energy at that time, the results demonstrate how bioeconomic models can be utilized to design aquacultural systems so that they optimize the use of energy (and possibly other inputs); thus bioeconomic modeling constitutes a viable tool for the development of sustainable aquacultural systems.

In research applications, bioeconomic models represent a powerful tool for the formulation, examination, and improvement of hypothesis and theories (Cuenco, 1989). In that regard, Allen and Johnston (1976) developed a bioeconomic model of an aquaculture facility with the ultimate purpose of demonstrating how these models could be used to identify key parameters in aquaculture production, give insight into cost-effective research, and allow for the investigation of economic feasibility.

In general, it is worthwhile noting that in all these applications, use of bioeconomic models results in the "optimal" allocation of resources. However, a word of caution is warranted. We should ask the question, Optimal in what sense? The primary objective of these models is maximizing profits or minimizing costs; i.e., these models seek the optimal allocation of resources from an economic perspective only. The recommendations of these models can thus be thought of as necessary but not sufficient for the development of sustainable aquacultural systems and management practices.

Applications of Bioeconomic Models at the Regional Level. There are few applications of bioeconomic models at the regional level in aquaculture. In fact, Sylvia and Anderson's (1993) study is the only one to date. These authors have developed a bioeconomic model for developing information for private and public salmon aquacultural policy strategies when environmental issues are important. The two levels of analysis refer to the two actors: salmon producers and policy makers. While the producers are assumed to maximize profits, the public policy makers are faced with four policy objectives: revenue, benthic quality, profits, and tax revenues. The policy instruments in their study include the number of allowable sites and an effluent tax. The two-level problem is not generally solvable in one step. However, an approximation can be obtained by using iterative simulation techniques and response functions.

MODELING SUSTAINABILITY

Sustainability is a popular concept in developmental planning. However, sustainability is a rather elusive concept and means different things to different groups. Thus we discuss it, once more, from the vantage point of modeling. There are many definitions of sustainability in the literature from various disciplines. Ecologists define a sustainable agricultural (aquacultural) system as one with a nonnegative change in the stock of natural resources and environmental quality over time. Economists define a sustainable agricultural (aquacultural) system as one with a nonnegative trend in total-factor social productivity (defined as the total value of all output produced by the system during one cycle, divided by the total value of all inputs used) (Lynam and Herdt, 1989). It is obvious that sustainability of aquatic food production can be described using both ecological and economic principles and possibly social concerns. Although a unifying definition integrating the interdependency of ecological, economic, and social issues is difficult, it is generally agreed that for an aquatic system to be sustainable it must have the opportunity to persist and be renewed without negative environmental effects. In other words, sustainability is a goal whereby an aquatic system can produce in ways that can be continued for generations to come. A sustainable aquatic farm must be economically viable, ecologically (environmentally) sound, and socially (and politically) acceptable. An environmentally sound farm that fails to provide an acceptable standard of living for the farm producer will not last long. Similarly, a highly profitable farm that heavily exploits and degrades the environment will be doomed to fail in the long run.

Sustainability as a MultiCriteria Concept

It is clear that a sustainable aquatic farm must achieve its economic, environmental, and social goals simultaneously. In aquacultural development planning, decision makers often have to make tradeoffs among these rather conflicting goals. These multi- rather than single-goal planning exercises require an extension of the traditional single-goal planning models to a multicriteria framework. It is particularly important as traditional cost-benefit analysis often fails to express ecological and environmental values in terms of monetary units or as externalities. In this context, *multicriteria decision-making* (MCDM) models are most appropriate and can be very useful to support decision making. While MCDM models have been widely used in agriculture for operational as well as strategic planning purposes (Romero and Rehman, 1989), there is only one application to aquaculture (Sylvia and Anderson, 1993).

Straightforward applications of MCDM models are not evident due to the complexities of the developmental planning exercise. In the following section we describe a decision support system that aims systematically to assist the decision maker or planner in making choices regarding the developmental planning of the aquaculture industry for a given region. A MCDM modeling framework forms the core of this decision support system. Later in the chapter we will illustrate an application of this system to analyze the tradeoffs among economic, environmental, and social goals through a case study in northern Egypt.

An Aquacultural Development Decision Support System

An agricultural development decision support system (ADDSS) has three main parts: a modeling component, a database component, and a dialog component (El-Gayar et al., 1994; El-Gayar, 1995). The modeling component is comprised of a model base containing all relevant models that are essentially multiple-objective in nature, and a model base management system with provisions for model specification, building, and execution. The database component is comprised of a database containing all relevant data and a database management system with provisions for data handling (e.g., adding, deleting, and locating, and modifying data entries). Finally, the dialog component provides a user interface to the other two primary components of the system.

The MCDM model representing the core of the modeling component seeks to identify the optimal allocation of resources and activity levels that strike an acceptable balance among the various decision attributes under consideration, subject to resource, market, and pollution constraints. In general, regional planning is has a multitude of issues and attributes involved, and regional planning for aquacultural development is no exception. Examples of regional development goals include producing human food, improving natural stocks, improving the standards of living, and increasing foreign-exchange earnings. In realizing such goals, the planner has to make decisions regarding the level of activities (decision variables) that would strike an acceptable balance among such often conflicting goals. Examples of such decision variables include what species to grow, what technology to use, how much to grow of each species, and/or what technology to use, etc. Moreover, the planner is constrained by the resources available in the region such as land, labor, water, etc., as well as other external constraints such as domestic market demand, export market demand, and pollution restraints. The MCDM model thus seeks to assist the planner in identifying feasible alternative decisions (i.e., those satisfying resource availability and external constraints) that attempt to reach a balance among the multiple goals and objectives encountered when planning for regional aquacultural development. Figure 7.3 depicts the general framework for the MCDM model developed.

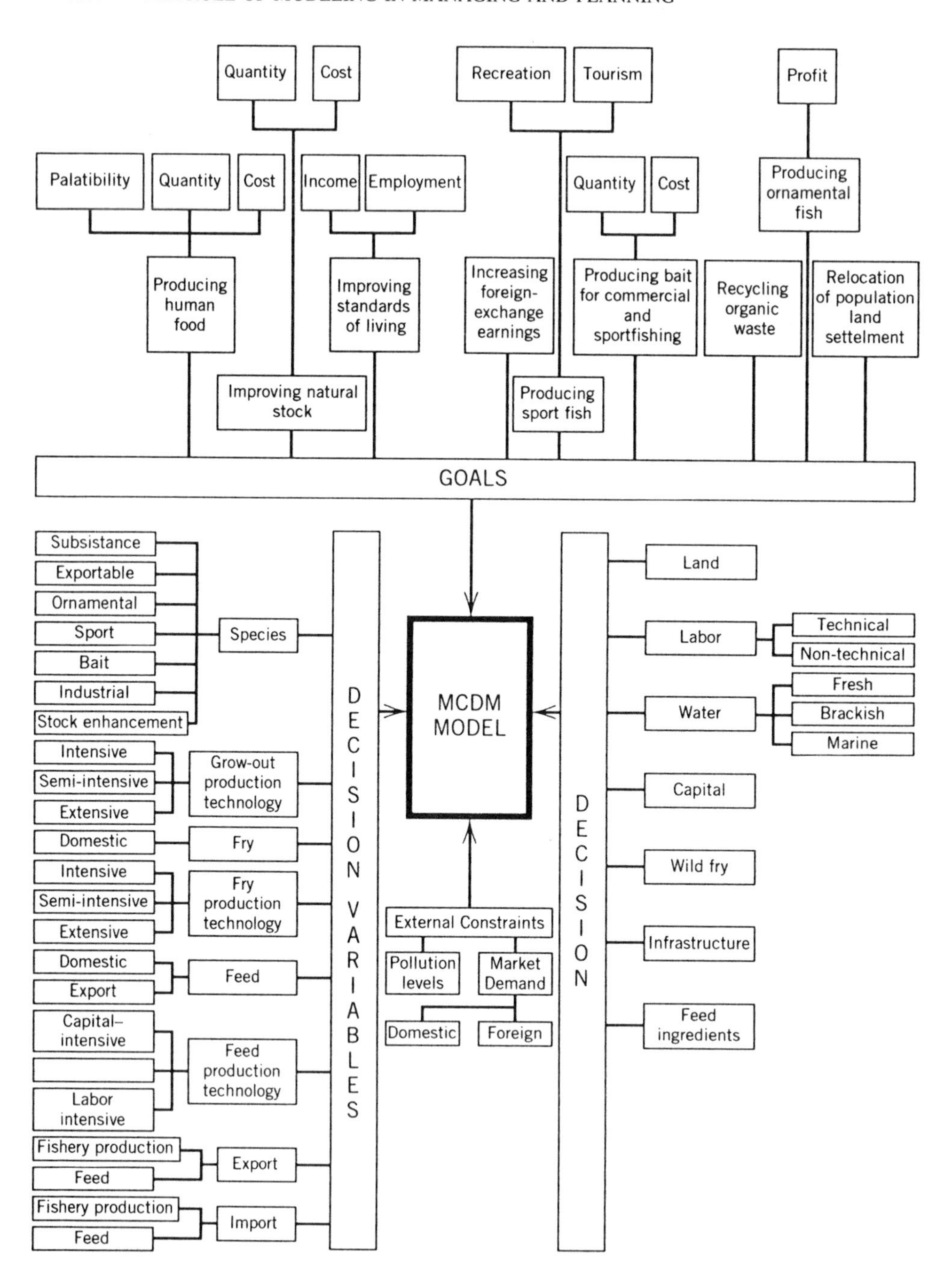

Figure 7.3. The multicriteria decision-making (MCDM) model.

To accommodate different decision situations, three different MCDM solution techniques are implemented: multiple-objective programming (MOP), compromise programming (CP), and weighted-goal programming (WGP). MOP is recommended in a decision situation that considers only two decision attributes. WGP is recommended when target values are known *a priori* for the decision attributes, while CP is recommended when the number of objectives exceeds 2 and target values are not known *a priori* for the decision attributes.

The system allows the planner to specify the model by selecting the decision attributes, the decision variables (activities), and the constraints to be considered. The user-specified model is generated using the data that is stored in the database and would answer questions such as, what species to grow? what technology to use? how much to grow of each species, and with what technology? The recommended policy is output in both text and graphics formats for ease of reference and analysis.

The dialog component of ADDSS is implemented as a graphical user interface (GUI) with a pull-down menu structure that operates under the Microsoft's Windows environment. Microsoft's FoxPro is used to provide all data management capabilities. GAMS (General Algebraic Modeling System) is used to develop and implement the MCDM model and techniques representing the core of ADDSS. Interested readers can refer to El-Gayar (1995) for details of the system's anatomy and the MCDM model's mathematical structure.

Case Study in Northern Egypt. The application of ADDSS is demonstrated through a case study from northern Egypt. This case study is primarily concerned with resolving issues pertaining to planning for aquacultural development in northern Egypt; here aquaculture appears to be a viable industry, with significant potential for supplying cheap and good-quality protein, helping to balance the foreign-exchange deficit, improving the standards of living, and creating employment opportunities. However, northern Egypt could accommodate a wide variety of aquaculture production systems and technologies. Such production systems could incorporate a variety of species, primarily sea breams, seabass, mullets, tilapias, and carps that are grown separately (monoculture) or in combinations (polyculture). Moreover, production systems could be employed at different technology levels ranging from intensive to semi-intensive to extensive culture systems. With such a range of systems and technologies, which vary in yield, product quality, profit, and utilization of resources, the choice of systems and the level of technology to employ would clearly affect the levels of the different decision attributes; the possible need to compromise between these attributes would thus complicate the decision-making process. It should also be noted that in any situation the choice of production systems and technologies is not the only controversial issue. Similar issues include the choice of fry and of feed production systems and technologies. Export and import decisions also play a role; of particular interest is identifying how much to import and how much to export from each species and feed type.

This case study considered four decision attributes: regional availability of protein, income, employment, and foreign-exchange earnings. The results indicate that ADDSS was successful here in tackling the case study by pointing out the conflict between the decision attributes, particularly regional availability of protein versus foreign-exchange earnings, and recommending alternative policies to the planner. Using MOP, three alternative policies were identified: a foreign-exchange policy that focuses on aquacultural development for the purpose of generating foreign-exchange earnings, a food security policy that focuses on increasing the regional availability of protein, and a compromise policy that strikes a balance between foreign-exchange earnings and regional availability of protein. Furthermore, in tackling the case study under consideration, it became evident that the flexibility and ease of use of ADDSS can allow the planner to explore the decision situation in a multitude of ways, thereby supporting him or her in making more informed decisions.

Tradeoff among Economic, Environmental, and Social Goals

Recognizing the importance of aquaculture on the environment, the ADDSS described earlier was modified to incorporate pollution as a decision attribute in the MCDM model instead of as an external constraint. This allowed the planner to explicitly address the tradeoffs among environmental and other economic and social goals. As an illustration, the tradeoff between the regional availability of protein and pollution was estimated using the modified ADDSS with the baseline information from the case study in northern Egypt as described earlier. Furthermore, a best-compromise solution is derived using compromise programming.

Figure 7.4 shows the estimated tradeoff between pollution and the regional availability of protein. Pollution was measured here as an index based on a scale of 1 to 10 per unit of production. Until actual pollution data are available, we have no recourse but to use a subjective index. As shown in Figure 7.4, at provision levels up to 22,000 tons of fish for regional use, pollution does not pose much of a problem. This is because, in this plan, all of this fish (relatively cheap carp) could be imported, thus exhausting the available foreign exchange reserves. However, from 22,000 to about 63,000 tons, there is an apparent tradeoff between pollution and providing fish for regional use. The tradeoff is more pronounced beyond 63,000 tons, as indicated by the steeper slope of the tradeoff curve. The actual values of the tradeoff (i.e., the opportunity costs) are represented by the slopes of the line segments connecting the efficient points.[1] For example, the line segment from 22,000 to 63,000 tons indicates that in this part of the tradeoff curve, each ton of increase in regional availability of fish increases the pollution index by about 10 units. In other words, the opportunity cost of increasing the regional availability of protein by 1 ton is to "sacrifice" the environment with an increased level of pollution by about 10 units. Beyond 63,000 tons, the opportunity cost increases to about 27 units of pollution. It is evident from the tradeoff curve that no more than 22,000 tons of fish

[1]An efficient point (combination of pollution level and the regional availability of protein) is defined such that it is not possible to move feasibly so as to increase one of the objectives without necessarily decreasing the other.

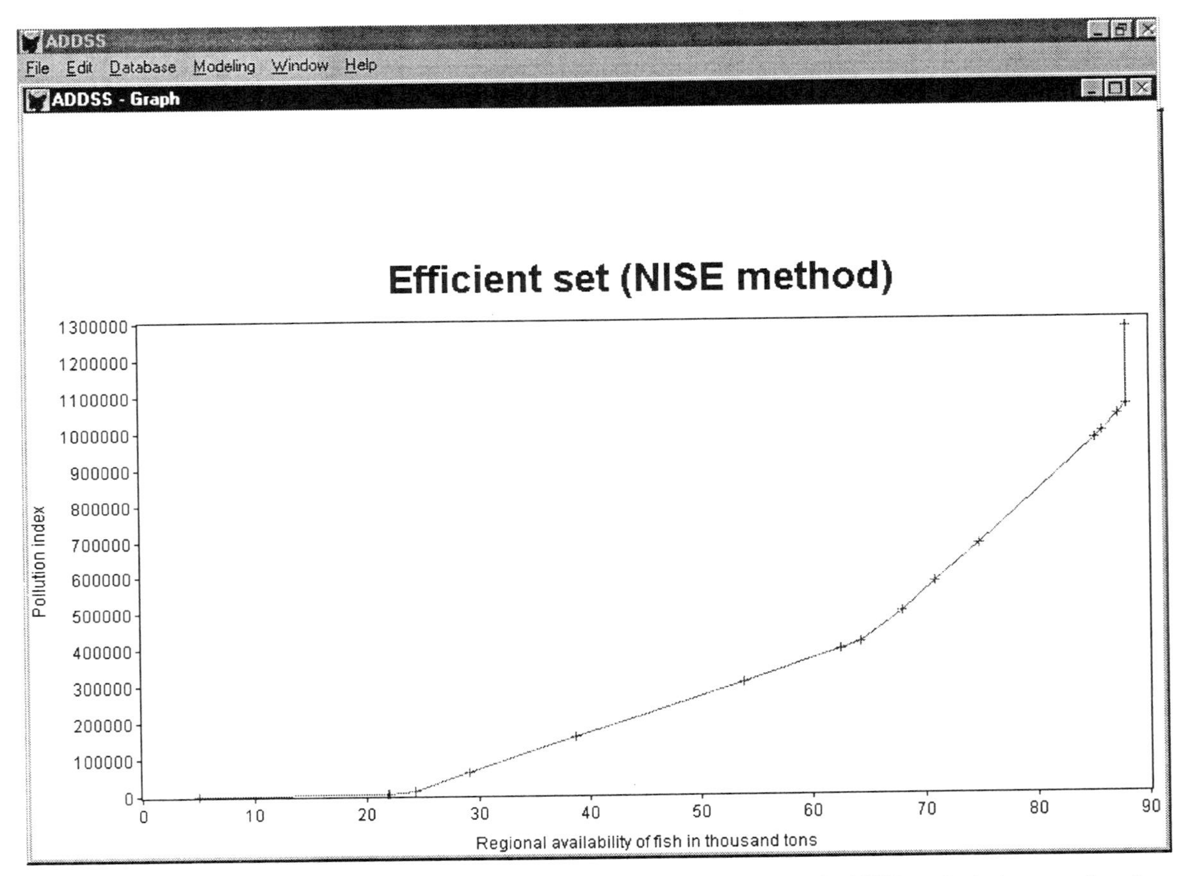

Figure 7.4. The efficient set for regional availability of protein and pollution using the NISE method: the case of northern Egypt.

can be made available for regional consumption if absolutely no pollution is desired. On the other hand, pollution would reach 1.1 million units if we wish to satisfy the entire demand of regional protein need. In general, the planner or decision maker will choose the best preferred plan from the set of efficient solutions as represented by the tradeoff curve based on his or her preferences regarding each of these two goals.

Without soliciting the exact preference of the decision maker, ADDSS also provides a best-compromise solution that can be thought of as the closest point to the ideal, i.e., to satisfying all regional fish demand with no pollution. Obviously, the ideal point is unattainable. In this case (using the infinity metric), the best-compromise point is found to satisfy about 63,000 tons of regional protein need (about 70% of ideal) with a pollution index of about 400,000 units. This corresponds to the point where the slope of the tradeoff curve changes from 10 units of pollution per ton of regional availability of fish, to 27 units of pollution. This compromise solution (plan) is outlined in Tables 7.2 to 7.4. The compromise solution suggests a balance of acreage allocated between production systems whose output is designated for domestic use (mullet) and as those whose output is aimed at satisfying export market demand (sea bream and seabass). The revenue from the export operation is used to import carp, thereby increasing the regional availability of protein. This balance approach keeps pollution to a more manageable level compared with other alternatives. In addition, because of the relatively high level of pollution associated with feed and fry production, the solution suggests importation of both feed and fry to support the aquaculture industry.

CONCLUDING REMARKS

This chapter reviewed the current status of modeling aquaculture production systems for operational management as well as strategic planning. It is evident that the traditional single-goal modeling techniques are inadequate in modeling and evaluating the sustainability of aquaculture production where multiple goals are sought. MCDM models provide a natural extension to the traditional single-goal techniques by addressing the multigoal nature of sustainable production and development.

It should be noted, however, that systems simulation of mechanistic models can also be helpful in predicting or evaluating the sustainability of aquaculture production systems at the farming system level. Conway (1985) defined sustainability as "the ability of a system to maintain productivity in spite of a major disturbance, such as is caused by intensive stress or a large perturbation." This definition is particularly useful when a mechanistic model of aquaculture production systems is available for systems simulation to predict whether the system can withstand major disturbances.

This chapter also presented a MCDM modeling system for aquacultural development planning in which multiple developmental goals could be addressed. It further illustrated how such a system can be used to analyze the tradeoffs among the conflicting goals of economic growth, social acceptability, and environmental

TABLE 7.2. Compromise Solution Generated from ADDSS (the Decision Variables)

Decision Variables	Culture Intensity	Compromise Solution
I. Systems whose output is designated for domestic consumption:		(in *acres*)
Sea brem monoculture	Intensive	0.24
	Semi-intensive	0
Seabass monoculture	Intensive	0
	Semi-intensive	0
Mullet monoculture	Semi-intensive	63,298
Sea bream/mullet/tilapia polyculture	Semi-intensive	0
	Extensive	0
Seabass/mullet/tilapia polyculture	Semi-intensive	0
	Extensive	0
Mullet/tilapia/carp polyculture	Semi-intensive	0
	Extensive	0
Seabass/sea bream/mullet/tilapia polyculture	Semi-intensive	0
	Extensive	0
II. Systems whose output is designated for export:		(in *acres*)
Sea bream monoculture	Intensive	625
	Semi-intensive	0
Seabass monoculture	Intensive	625
	Semi-intensive	0
Sea bream/mullet/tilapia polyculture	Semi-intensive	0
	Extensive	0
Seabass/mullet/tilapia polyculture	Semi-intensive	0
	Extensive	420
Seabass/sea bream/mullet/tilapia polyculture	Semi-intensive	0
	Extensive	0
III. Fish imported:		(in *tons*)
Carp		22,000
IV. Fry produced for supplying domestic aquaculture:		(in *million fry*)
Sea bream	Capital-intensive	2.21
	Labor-intensive	0
Seabass	Capital-intensive	2.54
	Labor-intensive	0
Mullet	Capital-intensive	167.86
	Labor intensive	
Tilapia	Capital-intensive	0.42
	Labor-intensive	0
Carp	Capital-intensive	0
	Labor-intensive	0
V. Fry produced for stocking natural reservoirs:		(in *million fry*)

(continued)

TABLE 7.2. (continued)

Decision Variables	Culture Intensity	Compromise Solution
Mullet	Capital-intensive	0
	Labor-intensive	0
VI. Feed produced for supplying domestic aquaculture:		(in *tons*)
A. Extensive polyculture	Capital-intensive	0
B. Intensive and semi-intensive polyculture	Capital-intensive	
C. Sea bream and seabass monoculture	Capital-intensive	
D. Mullet monoculture	Capital-intensive	
VII. Feed produced for export:		(in *tons*)
B. Intensive and semi-intensive polyculture	Capital-intensive	0
C. Sea bream and seabass monoculture	Capital-intensive	0
VIII. Feed imported:		(in *tons*)
B. Intensive and semi-intensive polyculture		516
C. Sea bream and seabass monoculture		5,780
D. Mullet monoculture		63,126
IX. Feed ingredients imported:		(in *tons*)
Corn		0
Soybean		0
Fish meal		0

TABLE 7.3. Compromise Solution Generated from ADDSS (the Resources)

Resource Name	Limit	Compromise Solution
Land:	(in acres)	
For freshwater aquaculture	10,000	323
For brackish-water aquaculture	200,000	63,298
For marine-water aquaculture	30,000	1,250
Labor:	(in man-hr × 10^6)	
Technical labor	150	2
Nontechnical labor	1,400	3
Water:	(in cubic metric tons/day × 10^6)	
Fresh water	1	0
Brackish water	37	25
Marine water	10	10
Capital:	(in L.E. × 10^6)[a]	
Domestic capital	1,000	433
Foreign reserves	10	10
Wild fry:	(in fry × 10^6)	
Sea bream	6	6
Seabass	6	6
Mullet	96	96
Feed ingredients:	(in tons)	
Rice bran	30,000	0

[a]L.E. = Egyptian pounds.

TABLE 7.4. Compromise Solution Generated from ADDSS (the Markets)

The Market	Demand (metric tons)	Compromise Solution (metric tons)
Domestic fish market		
Sea bream	7,500	0
Seabass	1,500	0.71
Mullet	45,863	40,511
Tilapia	32,100	0
Carp	22,000	22,000
Export fish market		
Sea bream	2,000	2,000
Seabass	2,000	1,894
Export feed market		
B. Intensive and semi-intensive polyculture	8,000	0
C. Sea bream and bass monoculture	8,000	0

soundness of various development plans and policies. While the MCDM framework provides a powerful tool to handle the multigoal aspects of developmental planning, it is useful only if it can easily be accessed by potential planners and policy decision makers. This chapter described a decision support system that integrates the MCDM framework into aquacultural development planning for a region.

It is obvious that sophistication in modeling software and the unprecedented computing power available today to modelers will increase the realism of aquacultural models to address complex issues such as sustainability. However, the ever-increasing number of aquacultural models as well as their increasing complexities emphasize the need for using decision support system technology that would bring aquacultural models from the theoretical world of academia into the real world of practical applications.

REFERENCES

Adams, C. M., W. L. Griffin, J. P. Nichols, and R. W. Bricks. 1980a. Bioengineering—economic model for shrimp mariculture system. TAMU-SG-80–203. Texas A&M University, 118 p.

Adams, C. M., W. L. Griffin, J. P. Nichols, and R. W. Bricks. 1980b. Applications of a bioeconomic engineering model for shrimp mariculture systems. South. J. Agric. Econ., 12: 135–141.

Allen, P. G., and W. E. Johnston. 1976. Research direction and economic feasibility: an example of systems analysis for lobster aquaculture. Aquaculture, 9: 155–180.

Allen, P. G., L. W. Botsford, A. M. Schuur, and W. E. Johnson. 1984. Bioeconomics of Aquaculture. Elsevier Science Publishers, New York.

Alsagoff, S. K., H. A. Clonts, and C. M. Jolly. 1990. An integrated poultry, multi-species aquaculture for Malaysian rice farmers: a mixed integer programming approach. Agric. Syst., 32: 207–231.

Alsagoff, S. K., K. Kuperan, and M. A. Hussein. 1992. A linear programming analysis of integrated agriculture-aquaculture mixed farming. Pertanika, 15(3): 271–278.

Azizan, Z. 1983. Using simulation methods to determine the optimal harvesting period for the cultured Malaysian freshwater prawns, Macrobrachium rosenbergii. Masters thesis, University of Illinois, Champaign-Urbana. 196 p.

Bacon, J. R., C. M. Gempesaw II, I. Supitaningsih, and J. Hankins. 1993. Risk management through integrating aquaculture with agriculture. pp. 99–111, in: J. K. Wang (Ed.). Techniques for modern aquaculture. American Society of Agricultural Engineers, St. Joseph, Minn.

Bala, B. K., and M. A. Satter. 1989. System dynamic simulation and optimization of aquacultural systems. Aquacult. Eng. 8: 381–391.

Bernard, D. R. 1983. A survey of the mathematical models pertinent to fish production and tropical pond aquaculture. pp. 225–235, In: J. E. Lannan, R. O. Smitherman, and G. Tchobanoglous. Principles and practices of pond aquaculture Oregon State University, Corvallis, Ore.

Bjorndal, T. 1988. Optimal harvesting of farmed fish. Mar. Resour. Econ., 5: 139–159.

Botsford, L. W. 1977. Current economic status of lobster culture research. Proc. World Maricult. Soc., 8: 723–739.

Botsford, L. W., H. E. Rauch, and R. A. Shleser. 1974. Optimal temperature control of a lobster plant. IEEE Trans. Automatic Control AC-19: 541–543.

Botsford, L. W., H. E. Rauch, A. A. Schuur, and R. A. Schleser. 1975. An economically optimum aquaculture facility. Proc. World Maricult. Soc., 6: 407–420.

Botsford, L. W., J. C. Van Olst, J. M. Carlberg, and T. W. Gossard. 1977. The use of mathematical modeling and simulation to evaluate aquaculture as a beneficial use of thermal effluent. pp. 405–410, In: Proceedings of the 1977 Summer Computer Simulation Conference, July 18–20 Chicago.

Cacho, O. J. 1993. Development and Implementation of a fish-farm bioeconomic model: a three-stage approach. pp. 57–72, In: U. Hatch and H. Kinnucan (Eds.). Aquaculture: models and economics. Westview Press, Boulder, Colo.

Cacho, O. J., U. Hatch, and H. Kinnucan. 1990. Bioeconomic analysis of fish growth: effects of dietary protein and ration size. Aquaculture, 88: 223–238.

Cacho, O. J., H. Kinnucan, and U. Hatch. 1991. Optimal control of fish growth. Am. J. Agric. Econ., 73: 174–183.

Callaghan, S. S. 1975. A dynamic model of a proposed heated finishing plant for oysters. M. S. thesis, University of Massachusetts, Amherst. 112 p.

Chamberlain, G., and H. Rosenthal. 1995. Aquaculture in the next century: opportunities for growth and challenges of sustainability. World Aquacult., 26(1): 21–25.

Clark, K. 1994. Optimal harvesting eco-economic model for integrated farming systems. Ph.D. dissertation, Department of Agricultural and Resource Economics, University of Hawaii, Honolulu. 201 p.

Conway, G. R. 1985. Agrosystem analysis. Agric. Admin., 20: 31–55.

Cuenco, M. L. 1989. Aquaculture systems modeling: an introduction with emphasis on warm water aquaculture. ICLARM Studies and Reviews 19. 46 p.

El-Gayar, O. F. 1995. An aquacultural development decision support system (ADDSS). Ph.D. dissertation, University of Hawaii at Manoa. 208 p.

El-Gayar, O. F., and Leung, P. S. 1994. A risk programming model for brackish water polyculture: a case study in Egypt. Paper presented at the 25th Conference of the World Aquaculture Society, January 1994, New Orleans.

El-Gayar, O. F., P. S. Leung, and L. Rowland. 1994. An aquacultural development decision support system (ADDSS): a preliminary design. Paper presented at the 25th Conference of the World Aquaculture Society, January 1994, New Orleans.

Emanuel, W. R., and R. J. Mulholland. 1975. Energy based dynamic model for Lago Pond, Ga. IEEE Trans. Automatic Control, AC-20: 98–101.

Engle, C. R., and U. Hatch. 1988. Economic assessment of alternative aquaculture aeration strategies. J. World Aquacult. Soc., 19: 85–96.

Engle, C. R., and G. Pounds. 1993. Economics of single- and multiple-batch production regimes for catfish. pp. 75–89, In: U. Hatch and H. Kinnucan (Eds.). Aquaculture: models and economics. Westview Press, Boulder, Colo.

Faustmann, M. 1849. On the determination of the value which forest land and immature stands possess for forestry. pp. 441–455, In: M. Gane (Ed.). Martin Faustmann and the evolution of discounted cash flow. Commonwealth Forestry Institute Paper 42. Oxford, England.

Gates, J. M., C. R. MacDonald, and B. J. Pollard. 1980a. Salmon culture in water reuse system: an economic analysis. University of Rhode Island Marine Technical Report #78. Kingston, Rhode Island.

Gates, J. M., C. R. MacDonald, and B. J. Pollard. 1980b. A dynamic linear programming model of fish culture in water reuse systems. Rhode Island Agricultural Experiment Station Contribution 1892. 20 p.

Gempesaw II, C. M., F. F. Wirth, J. R. Bacon, and L. Munasinghe. 1993. Economics of vertical integration in hybrid striped bass aquaculture. pp. 91–105, In: U. Hatch and H. Kinnucan (Eds.). Aquaculture: models and economics. Westview Press, Boulder, Colo.

Griffin, W. L., J. S. Hanson, R. W. Brick, and M. A. Johns. 1981. Bioeconomic modeling with stochastic elements in shrimp culture. J. World Maricult. Soc., 12: 94–103.

Griffin, W. L., L. A. Jensen, and C. M. Adams. 1983. A generalized budget simulation model for aquaculture. TAMU-SG-83-202. Texas A&M University, College Station. 131 p.

Griffin, W. L., W. E. Grant, R. W. Brick, and J. S. Hanson. 1984. A bioeconomic model of shrimp maricultural systems in the USA. Ecol. Model., 25: 47–68.

Hanson, J. S., W. L. Griffin, J. W. Richardson, and C. J. Nixon. 1985. Economic feasibility of shrimp farming in Texas: an investment analysis for semi-intensive pond grow-out. J. World Maricult. Soc., 16: 129–150.

Hatch, U., and J. Atwood. 1988. A risk programming model for farm-raised catfish. Aquaculture, 70: 219–230.

Hatch, U., and Kinnucan, H. 1993. Introduction. pp. 1–13, In: U. Hatch and H. Kinnucan (Eds.). Aquaculture: models and economics Westview Press, Boulder.

Hatch, U., S. Sindelar, D. Rouse, and H. Perez. 1987. Demonstrating the use of risk programming for aquacultural farm management: the case of penaeid shrimp in Panama. J. World Aquacult. Soc., 18: 260–269.

Hatch, U., J. Atwood, and J. Segar. 1989. An application of safety-first probability limits in a discrete stochastic farm management programming model. South. J. Agric. Econ., 18: 65–72.

Hazell, P. B. R., and Norton, R. D. 1986. Mathematical programming for economic analysis in agriculture. Macmillan, New York.

Hean, R. L. 1994. An optimal management model for intensive aquaculture—an application in Atlantic salmon. Aust. J. Agric. Econ., 38(1): 31–47.

Hochman, E., P. S. Leung, L. W. Rowland, and J. A. Wyban. 1990. Optimal scheduling in shrimp mariculture: a stochastic growing inventory problem. Am. J. Agric. Econ., 72: 382–393.

Johnson, F. C. 1974. Hatch—a model for fish hatchery analysis. Report NBSIR 74-521. National Bureau of Standards, Washington, D.C. 51 p.

Johnston, W. E., and L. W. Botsford. 1981. Systems analysis for lobster aquaculture. pp. 455–464, In: Proceedings of the World Symposium on Aquaculture in Heated Effluents and Recirculation Systems, Vol. II. May 28–30, 1980, Stavanger, Berlin.

Karp, L., A. Sadeh, and W. L. Griffin. 1986. Cycles in agricultural production: the case of aquaculture. Am. J. Agric. Econ., 68: 553–561.

LaFranchi, C. 1992. Optimal selection of species and harvest scheduling for cultured shrimp. Master thesis, Department of Agricultural and Resource Economics, University of Hawaii, Honolulu. 66 p.

Leung, P. S. 1993. Bioeconomic modeling in aquaculture after two decades. pp. 115–137, In: Y. C. Shang, P. S. Leung, C. S. Lee, M. S. Su, and I. C. Liao (Eds.). Proceedings of the International Symposium on Socioeconomics of Aquaculture. Taiwan: Taiwan Fisheries Research Institute, Taiwan.

Leung, P. S., and L. W. Rowland. 1989. Financial analysis of shrimp production: an electronic spreadsheet model. Comput. Electron. Agric., 3: 287–304.

Leung, P. S., and Y. C. Shang. 1989. Modeling prawn production management system: a dynamic Markov decision approach. Agric. Syst., 29: 5–20.

Leung P. S., Y. C. Shang, and X. Tian. 1994. Optimal harvest age for giant clam Tridacna derasa: an economic analysis. J. Appl. Aquacult., 4: 49–64.

Lipschultz, F., and G. E. Krantz. 1980. Production optimization and economic analysis of an oyster (Crassostrea virginica) hatchery on the Chesapeake Bay, Maryland, USA. Proc. World Maricult. Soc., 11: 580–591.

Logan, S. H., and W. E. Johnston. 1993. A replacement model for rainbow trout broodstock under photoperiod control. pp. 107–122, In: U. Hatch and H. Kinnucan (Eds.). Aquaculture: Models and Economics Westview Press, Boulder, Colo.

Logan, S. H., and K. Shigekawa. 1986. Commercial production of sturgeon: the economic dimensions of size and product mix. Giannini Research Report 335. University of California, Davis. 69 p.

Lynam, J. K., and R. W. Herdt. 1989. Senses and sustainability: sustainability as an objective in international agricultural research. Agric. Econ., 3: 381–398.

McNown, W., and A. Seireg. 1983. Computer aided optimum design and control of staged aquaculture systems. J. World Maricult. Soc., 14: 417–433.

Pardy, C. R., W. L. Griffin, M. A. Johns, and A. L. Lawrence. 1983. A preliminary economic analysis of stocking strategies for penaeid shrimp culture. J. World Maricul. Soc., 14: 49–63.

Piedrahita, R. H. 1988. Introduction of computer modeling of aquaculture pond ecosystems. Aquacult. Fish. Manage., 19: 1–12.

Rauch, H. E., L. W. Botsford, and R. A. Shleser. 1975. Economic optimization of an aquaculture facility. IEEE Trans. Automatic Control, AC-20: 310–319.

Romero, T., and T. Rehman. 1989. Multiple criteria analysis for agricultural decision. Elsevier Science Publishers, Amsterdam.

Sadeh, A. 1986. Value of information in production processes: the case of aquaculture. Ph.D. dissertation, Texas A&M University, College Station. 137 p.

Sadeh, A., H. Talpaz, D. Bessler, and W. L. Griffin. 1989. Optimization of management plans with short and long run problems: The case of shrimp production. Eur. J. Operational Res. 40: 22–31.

Schuur, A. M., P. G. Allen, and L. W. Botsford. 1974. An analysis of three facilities for the commercial production of *Homarus americanus*. University of California, Bodega Marine Laboratory, Bodega Bay, Calif. 19 p.

Shaftel, T. L., and B. M. Wilson. 1990. A mixed-integer linear programming decision model for aquaculture. Managerial Decision Econ., 11: 31–38.

Sparre, P. 1976. A Markovian decision process applied to optimization of production planning in fish farming. Meddr Danm. Fisk.-og Havunders., 7: 111–197.

Springborn, R. R., A. L. Jensen, W. Y. B. Chang, and C. Engle. 1992. Optimum harvest time in aquaculture: an application of economic principles to a Nile tilapia, *Oreochromis niloticus*, growth model. Aquacult. Fish. Manage., 23: 639–647.

Summerfelt, S. T., J. A. Hankins, S. R. Summerfelt, and J.M. Heinen. 1993. Modeling continuous culture with periodic stocking and selective harvesting to measure the effect on productivity and biomass capacity of fish culture systems. In: J. K. Wang (Ed.). Techniques for modern aquaculture. Amer. Soc. of Agric. Engineers. St. Joseph, Minn. 581–595.

Syed, A. L. A. 1985. A mixed integer programming approach for integrating poultry with multi-species aquaculture on Malaysian rice farms. Ph.D. dissertation, Auburn University, Auburn, Ala. 166 p.

Sylvia, G., and J. L. Anderson. 1993. An economic policy model for net-pen salmon farming. pp. 17–38, In: U. Hatch and H. Kinnucan (Eds.). Aquaculture: Models and Economics Westview Press, Boulder, Colo.

Talpaz, H., and Y. Tsur. 1982. Optimizing aquaculture management of a single-species with fish population. Agric. Syst., 9: 127–142.

Tauer, L. W. 1983. Target MOTAD. Am. J. Agric. Econ., 65: 606–610.

Tian, X. 1993. Optimal aquafarm structure and size: a case study of shrimp mariculture. Ph.D. dissertation, Department of Agricultural and Resource Economics, University of Hawaii, Honolulu. 179 p.

Tisdell, C. A., L. Tacconi, J. R. Barker, and J. S. Lucas. 1993. Economics of ocean culture of giant clams, *Tridacna gigas*: internal rate of return analysis. Aquaculture, 110: 13–26.

Tsur, Y., and E. Hochman. 1986. Economic aspects of the management of algal production. pp. 473–483, In: CRC handbook of microalgal mass culture. CRC Press, Boca Raton, Fla.

Wu, R. S. S. 1989. Biological and economic factors in the selection of cultured fish species and the development of a bioeconomic model. AQUACOP IFREMER Actes Colloq., 9: 437–444.

8 Sustainable Aquaculture and Integrated Coastal Management

CHUA THIA-ENG

INTRODUCTION

The term *aquaculture* as used here denotes the farming of aquatic organisms for food or for commercial purposes. Aquacultural produce used to come primarily from fresh water. In recent years, however, increasing aquaculture production has occurred in coastal areas and on the open seas. Thus, the term *coastal aquaculture* denotes the cultivation of brackish-water or marine species in the ponds, protected coves, bays, gulfs, and lagoons. The term *mariculture*, on the other hand, is often used to denote the farming of marine organisms in the open seas, either in the water column or along the seabed.

Despite Asia's long history of development, aquaculture became an important commercial food production system and a significant source of export earnings only in the late 1960s (Chua, 1986, 1994). It started as a small-scale, labor-intensive, agrobusiness-based industry, producing food fishes mainly for local consumption. Species cultured were mostly low on the food chain, including the various carps and tilapias in inland waters or oysters and mussels in the coastal waters. These were normally farmed in earthen ponds or in open coastal waters.

Then, in the late 1970s, aquaculture transformed rapidly into a large-scale, high-technology, food-producing, and export-earning industry. Species high on the food chain were mostly cultured as a result of improved farming technology, successes in fish breeding, artificial feed development, and genetic improvements. After World War II, considerable economic achievements in many developing nations gave rise to the improvement of living standards; this in turn led to a higher demand for and greater choice in high-priced commodities, such as shrimps, salmons, eels, yellow-tails, sea breams, groupers, and seabasses (Chua, 1994).

The increasing recognition of aquaculture as an important source of fish supplies, employment for rural areas, and export earnings created a conducive policy environment and economic opportunity for large-scale development of shrimp and fish farms, leading to the intensification and expansion of the fish culture systems.

Sustainable Aquaculture, Edited by John E. Bardach
ISBN 0-471-14829-6

Encouraged by these spectacular, short-term economic gains, aquaculture was further supported by bilateral and multilateral lending institutions, especially the Asian Development Bank and the World Bank (Pauly and Chua, 1988). This support provided a greater investment opportunity for many developing nations. Coastal aquaculture, in particular shrimp farming, spread far and wide; within three decades, it emerged as a prime export-earning industry in many parts of the world (Liao, 1990).

Aquacultural development, however, also has its downside. In the last two decades, many intensive aquaculture enterprises have suffered severe losses due to disease outbreak (ADB and NACA, 1991), poor farm management, and the degrading quality of the aquatic environment (Chua and Paw, 1987; Chua, 1990, 1992, 1994). Shrimp farming, for instance, began to show signs of unsustainability in many countries where it had been intensified. More than U.S. $1 billion have been lost, mostly due to outbreaks of shrimp diseases in Asia (FAO and NACA, 1995). Disease outbreaks have been reported in Taiwan, Thailand, the Philippines, China, and a number of other nations. Many shrimp farming enterprises have collapsed due to severe drops in production, high operational costs, and fluctuating markets (Chua, 1994). With the acceleration in industrialization, intensified aquacultural development, urbanization, and dense population increases, the coastal waters where practically all maricultural activities were located have become increasingly polluted with sewage, and industrial and agricultural wastes, making many rivers, estuaries, bays, and lagoons unsuitable for fish farming (Chua et al., 1989; GESAMP, 1992). In 1978 to 1991 in Mansan-Chinhae Bay of South Korea, the occurrence of red tides was attributed to uncontrolled sewage discharge and dense maricultural activities, resulting in economic loss of U.S. $6.75 million. In northern China, mass mortalities of the penaeid shrimp *Penaeus orientalis* and mollusks have been attributed to the discharge of industrial wastes from paper mills (Wu, 1991), while in Taiwan, oysters have been found to be contaminated by the discharge of copper compounds from a recycling factory (Han and Huang, 1990).

The polluted waters have seriously affected the very life-support system essential for the cultivation of fish, shrimp, and other aquatic products. And due to the increasing competition for the limited resources, aquacultural development has inevitably run into conflicts with other uses (Chua, 1992), such as fishing, navigation, and port development. Finally, aquacultural practices that were inappropriately planned and managed have created other adverse environmental and socioeconomic impacts, in terms of habitat destruction, depletion, contamination of underground water, and threats to human health (Chua et al., 1989, GESAMP, 1992; ADB and NACA, 1995; Boromthanarat, 1995). In Panama, the reported rate of areal loss of mangroves is 1% a year, while in Ecuador, 34% of its 177,000 hectares of mangroves were converted to shrimp farms (Goldberg 1994). In the Philippines, Indonesia, and Thailand, large areas of mangroves have been converted to shrimp farms (Paw and Chua, 1991). The decline in cage culture operation in the lakes and lagoons in the Philippines is an excellent example of self-pollution caused by unplanned and unregulated aquacultural development.

As a result, the development of coastal aquaculture has shown signs of gradual decline in such countries as Korea (Kim, 1992), Taiwan (Chien, 1992), Japan (Miki and Sano, 1992), and Thailand (Tokrisna, 1992).

On the other hand, the existing fish catch of about 100 million metric tons (MMT) is not sufficient to meet the ever-increasing demand for fish supplies (Chua, 1994), which is expected to increase even faster in the next decade when the harvests from lakes, seas, and oceans reach their maximum sustainable level. And so there is currently a shortfall of about 30 MMT, which have to be provided through aquaculture (FAO, 1991).

The question of sustainability, therefore, has increasingly been asked. Can the aquaculture industry be sustained in its present form? Does it have the ability to sustain the socioeconomic benefits and to optimize the level of its activities?

Sustainability has become a major challenge to aquacultural development, which in many countries has reached a crossroads. Further development requires a reorientation not only in operation and management at the farm level but also greater control, integrated planning, and management of the industry by the state.

This chapter focuses on the needs and ways to manage the aquaculture industry in the overall context of integrated coastal management. The recommended approach and management strategies are largely based on the socioeconomic and environmental conditions in most developing countries in Asia, a region where most aquaculture now occurs and where the demand for fish protein will remain high in the future.

FACTORS AFFECTING SUSTAINABLE AQUACULTURAL DEVELOPMENT

The sustainability of an aquaculture enterprise depends on the combined effects of two sets of factors—intrinsic and extrinsic. The intrinsic factors include water quality, culture techniques, location and operation of facilities, seed supply, species characteristics, and availability of artificial or natural feeds. The extrinsic factors refer to national policy, natural hazards, climate change, pollution, the market, introduction of exotic species, sociocultural conditions of the culture sites, and legislative control. All of these factors affect the culture operations, marketability of products, distribution of economic and social benefits, and the functional integrity of the resource systems.

The intrinsic factors may be addressed through adequate on-farm planning and management, which reduce the adverse impacts on the environment and sustain aquaculture production targets (Phillips, 1995). Most of the environmental problems arising from farming practices may be addressed within the aquaculture enterprise, i.e., the fish growers and the supporting industries.

Many, if not all, of the extrinsic or off-farm factors, however, require appropriate governmental policy and management interventions, in terms of resource or area allocation, land- and water-use planning, technical assistance, information services, legislative controls, and coordination with other related industries. The combined effects of intrinsic and extrinsic factors make it necessary to manage aquaculture as an integral part of the whole resource system on which it relies.

INTEGRATED COASTAL MANAGEMENT

Water and land are finite resources. They are part of the resource systems that generate goods and services. Most resource systems are capable of sustaining more than one economic activity, but their use must be carefully planned and managed in order to optimize their net social and economic benefits. This is especially important in the coastal areas, where a large proportion of the world population now resides and is undertaking a wide variety of economic activities (Chua and Garces, 1994; Tokrisna, 1995). Thus, the primary purpose of *integrated coastal management* (ICM) is to achieve the following sustainable development goals for the coastal areas:

- ensuring the rational use of natural resources
- minimizing conflicts over use
- protecting and preserving the functional integrity of the coastal ecosystems
- distributing the derived economic benefits more equitably

These goals are expected to be reached through systematic, coordinated planning and program implementation. This process includes identifying which management issues threaten sustainability and addressing these issues collectively through policy, management, and technological interventions (Figure 8.1).

The lack of national policies in many coastal nations has limited their development planning. The use of the coastal waters and open seas has usually been left out of the conventional process of national planning and development. Whenever it has been included, sea-use planning was restricted to ports and harbors. Thus, application of the ICM process will enable the inclusion of the coastal waters and the open seas within their jurisdiction in future national development plans (Vallejo, 1993). A national policy not only contributes to the avoidance of use conflicts but also generates investment opportunities for the new economic activities in the coastal waters and the open seas.

ICM planning makes use of secondary information and rapid coastal appraisal for environmental profiling in order to identify management issues, information gaps, opportunities, and constraints (Pido and Chua, 1992). Through various research processes and a series of stakeholders' consultations, long-term strategic management plans and issue-oriented or special-area action plans are formulated.

The strength of ICM lies in its integration and coordination (Chua, 1993). It provides a holistic approach to addressing a host of complex environmental management problems in the coastal areas. It emphasizes and ensures policy and functional integration in the formulation and implementation of various specific action plans or activities. Policy conflicts often occur between central and local governments. Local ordinances, administrative orders, and legislation sometimes are not in

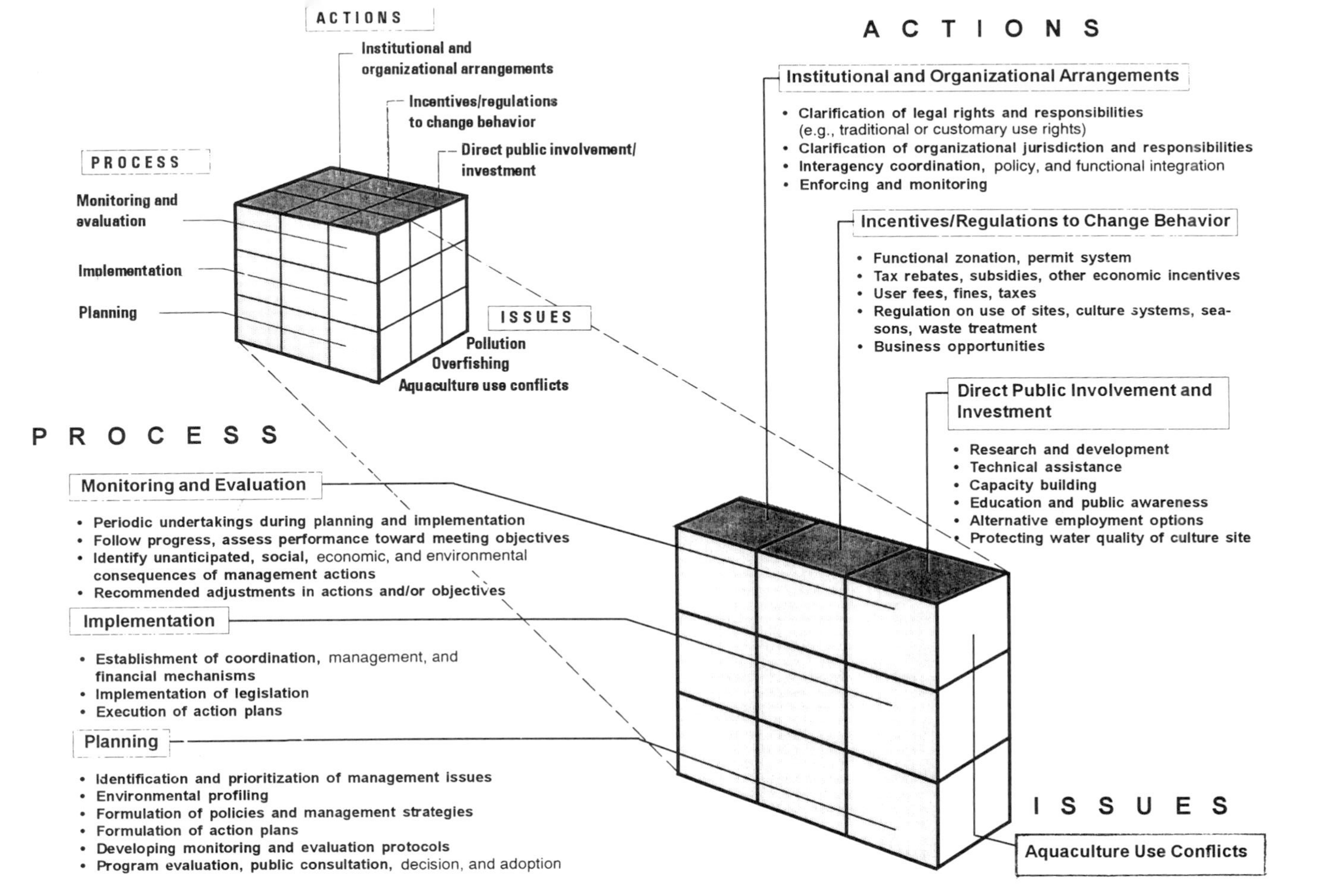

Figure 8.1. Application of integrated coastal management (ICM) framework to reduce conflicts between aquacultural development and other users.

harmony with the national legislation. Similarly, overlapping functions among the line agencies are not uncommon, often resulting in interagency turf conflicts.

ICM promotes the establishment of a coordinating mechanism that will execute the various action plans, mostly using the existing line agencies. It enables the concerned line agencies and other stakeholders involved in the governance, development, and use of the coastal and marine resources to plan collectively and implement agreed-on environmental management programs. This coordinating mechanism consists of a policy body and an implementing arm. The policy body usually takes the form of a commission (e.g., the Lingayen Gulf Coastal Resource Management Commission, Philippines; NEDA, 1992), a council (e.g., the proposed Batangas Bay Environmental Management Council), or a committee (e.g., the Integrated Marine Management and Coordination Committee of Xiamen Municipality, China). The policy body is made up of senior officials from the various line agencies, nongovernmental organizations (NGOs), and scientific organizations and representatives from stakeholders. This body should be functional and effective to address the various management issues. Thus, its composition should include senior officials who could devote the time and attention to implement the program. The Lingayen Gulf Commission is composed of the secretaries of concerned departments and the heads of relevant line agencies. However, finding a common time for the commissioners to meet regularly is already an impossible logistical task. As much as possible, therefore, this policy body should have been comprised of local officials who had genuine interests in the area in which they serve or know about instead of those coming from the central office or other regions.

The coordinating office may vary, depending on its management focus. Usually, a neutral line agency, such as the planning department, or a specialized agency, such as the Environment and Natural Resources Department (e.g., Batangas Province, Philippines), the Maritime Department (e.g., Hainan, China), or the Fisheries Department (e.g., Segara Anakan, Indonesia), may take the coordinating role. In some cases, a special coastal management office (e.g., Coastal Resources Planning and Management Unit, Johore, Malaysia) is created to implement ICM programs.

The implementation of specific action plans is normally carried out by the respective line agencies. It is crucial in the planning stage that the relevant line agencies be included in the plan formulation. The coordinating office should not be involved in the implementation of activities that a line agency can do. It is a basic principle of ICM that line agencies involved in this exercise will be strengthened and not be depleted of authority or financial resources.

Institutional coordination is one of the most important steps to ensure the implementation of policies, strategies, and action plans. In many cases, a great amount of effort is directed at formulation and research, that results in strategies and plans that are beautifully presented but lack the institutional and legal mechanisms needed to coordinate the implementation. The formulation of plans may take as long as 3 to 5 years, as in the case of the ASEAN/U.S. Coastal Resource Management Project (Scura et al., 1992). And it may take more than 5 years to get such a mechanism finally established (e.g., South Johore, Malaysia).

Another significant feature of ICM is the allocation and management of natural resources for specific uses. The area or resource allocation may be based on various criteria, such as the characteristics of the natural resources, traditional use rights, and national policies. The outcome of the resource allocation is the delineation of functional zones in which specific economic activities are encouraged and protected by legal measures. Thus, a specific zone may be allocated primarily for tourism development based on its potential, while another zone may be designed for all forms of industries after considering all the possible environmental impacts. An aquacultural development zone will be used mainly for aquaculture and aquaculture-friendly activities. In these zones, no other activities that are harmful to aquaculture, such as the discharges from sewage outfall or industrial wastes, will be allowed by law.

While ICM promotes public awareness on the importance of, and need for, a clean and safe coastal and marine environment, command and control measures are also instituted to ensure strict compliance with the laws and regulations. However, economic incentives are also needed to change human behavior. Such incentives may take the form of tax rebates, extended time for compliance to discharge standards, subsidies, and so on.

One of the most important components of ICM is the development of an integrated monitoring system that periodically reports on the status and trends of environmental quality as a result of the combined effects of various policy and management interventions.

ICM is usually regarded as a governmental program; hence it should be implemented within the existing or refined governmental framework. The roles of both central and local governments are defined in an ICM program. Improving public awareness and concerns about the environment is a public-sector responsibility, as are capacity building, research, and technical assistance. Hence, there is a need to strengthen governmental institutions in order to enable them to implement the ICM program.

An ICM program is most effective if developed for, and implemented at, the local level within a defined geographic area. The more local the administrative unit, the more effective the program integration and coordination will be. The local administrative structure has the greater ability to discern potential problems and can respond more promptly and effectively in a preventive, rather than reactive, management mode (Van Houtte, 1995). Where reactive management measures are required, the local administrative structure can fine-tune the application of national legislation, always aware of the sociocultural conditions, thus facilitating law enforcement. For a large country, a national ICM program is difficult to implement. In such a situation, a national policy that enables the local governments of coastal provinces and municipalities to develop their own ICM programs would be more helpful.

Although the ICM program has a general coverage of all economic activities, it does not preclude the necessity of focusing on the desired sectors or on specific environmental concerns. As such, it is possible to develop an ICM program focusing on aquaculture, biodiversity, fisheries, marine pollution, or coastal tourism, depending on the primary activities of the given area and the severity of the environmental concerns.

ICM is not a development program, nor does it provide activities or guidelines on how to develop a certain economic sector. On the contrary, it provides the guidelines and management measures necessary to prevent, control, or mitigate the adverse environmental impacts of economic activities. In fact, a well-structured ICM program provides the needed information for a comprehensive integrated environmental impact assessment and monitoring.

ICM includes an integrated assessment of the cumulative environmental impacts arising from existing and future economic activities. It performs a comprehensive environmental audit in relation to the ecological and socioeconomic conditions. It develops management strategies and action plans to neutralize the existing cumulative, adverse environmental impacts and to formulate preventive measures for future economic activities.

A good ICM program builds on the recommendations of the Integrated Environmental Impact Assessment (IEAS). An IEAS is still needed for individual, large-scale, economic projects, but the cost of undertaking an IEAS is greatly reduced due to the availability of baseline information and the general assessment already undertaken through the ICM program.

APPLICATION OF ICM FOR SUSTAINABLE AQUACULTURE DEVELOPMENT

Since the adoption of Agenda 21 at the U.N. Conference on Environment and Development (UNCED) in 1992, many countries have taken the initiative to adopt the ICM system for managing coastal and marine resources. There is also a major reorientation among the aid programs and bilateral and multilateral lending institutions to focus on developing ICM programs in coastal nations. However, despite the focus and attention given, many nationally and internationally funded projects on ICM are still looking for a generic ICM program framework that could be adopted in most coastal states. There are, in fact, limited examples of success stories, partly because of the long duration required to prove the program's effectiveness, but mainly because of the lack of expertise and experience in the formulation and execution of such programs.

The following are general descriptions of the approaches and guidelines for the applying the ICM framework, with a focus on sustainable aquacultural development.

Policy Environment

It is essential to create a policy environment that

a. enhances the willingness and cooperation of the industry and the fish farmers to undertake measures to reduce self-pollution
b. encourages local governments to undertake adequate aquacultural development planning and management
c. promotes the development of appropriate legislation and efficient law enforcement

A national policy on aquaculture reflects the government's commitment and support to a long-term development of the industry. Such a policy environment will strengthen the industry's willingness and cooperation in ensuring a healthy culture for their long-term investment. It is the general attitude and perception of investors that aquaculture, particularly shrimp culture, is undertaken to make a profit (Fegan, 1995). The protection of the environment is not their responsibility, and often a shotgun approach is to make as much profit as possible within the shortest possible time. Even among small fish farmers, the environmental protection of the culture sites is deemed to be the responsibility of the government.

Aquacultural development is highly localized, depending on its potential in a coastal area. The local government's involvement in the planning and development of the aquaculture industry ensures its continue support. The local government must understand the economic values and social benefits of aquaculture and commit to its long-term development. It is also important to realize that the failure to address aquacultural development issues at the local level could result in socioeconomic complications and instability in the rural areas. The failure of the cockle-culture industry in Penang, Malaysia, is a good example.

A favorable policy environment for aquacultural development provides substantive guidance on environmental protection and environmental legislation as well as economic incentives for aquacultural development. The industry must be convinced of the long-term gains for compliance and the penalties for noncompliance with the government regulations.

No policy makers will object to the need for sustainable development and for protecting the coastal and marine environment. However, the pressure for to eradicate poverty, the demand for rapid economic development, and political interest often make it difficult for policy makers to adhere to the program with a long-term goal. They tend to settle for a short-term program that brings about quick political and/or economic benefits, even at the expense of compromising the functional integrity of valuable ecosystems.

The challenge, therefore, is to present to the policy makers a menu of actions that harmonize environmental management with political and economic concerns. Policy makers will be convinced if they can see that accrued socioeconomic gains eventually contribute to their political benefit.

Efforts should be made to create national policies in favor of the application of ICM for aquacultural development at the local level. ICM also paves the way for the necessary institutional arrangements, which include capacity strengthening, associated budgetary allocations, and the transfer of authority from the central to the local governments.

Holistic and Integrated Planning and Management Framework

While aquacultural management is the main focus, the planning process should also consider the other users of the resources. First and foremost, the concerned fishery agency should be able to justify the allocation of the coastal areas for development, based on the ecological advantages and potential socioeconomic benefits and how these relate to the national and local government policies. There is also a need to identify other existing and potential users of the resources concerned and sort out existing or potential conflicts with all the affected sectors. This approach calls for the participation of other managers and government planners to determine jointly the best use of the resources that really reflect national priority and goals.

The conventional sectoral planning approach pertaining to the exploitation and use of limited natural resources has resulted in unhealthy spatial and temporal competition. Many ecologically important habitats, such as the coral reefs, mangroves, and seagrass beds, have been removed from other uses without considering the long-term adverse impacts on the fisheries these habitats sustain. In Xiamen, China, for instance, one of the world's few remaining habitats for *Amphioxus* (a living fossil) was rapidly being depleted because of the increased navigational activities, coastal constructions, and unregulated fishing intensity.

Initiating and Implementing Institutions

Most countries do not have specialized agencies to promote or establish ICM programs. Many existing ICM programs are being developed or executed by interested line agencies. It is appropriate for a line agency to initiate an ICM framework that focuses on the management and development of its area of responsibilities. Thus, a fishery agency may initiate an ICM framework to develop and manage the aquaculture enterprise. The ICM framework mandates the involvement or participation of other stakeholders in the planning and management of the resources. In situations where there are several prominent uses, including fishing, tourism, navigation, and the establishment and operation of a port or harbor, a neutral agency such as the economic commission or regional planning department will be more appropriate to take the lead. In the case of Hawaii, which is relatively developed but geographically homogenized and administratively unitary, the ICM program is lodged in the Office of State Planning and placed directly under the governor, although the state is attempting to move it to the Department of Land and Natural Resources.

In areas where aquaculture is the primary economic activity, the concerned fishery agency should take the leadership role in forging an integrated plan with the other sectors on the use of the resources. This is to ensure that the interest of the fish farmers and the industry's future development will be sustained and, at the same time, the potential conflicts with competing users will be minimized. It is expected that the fishery agency will play a more prominent and active, if not the leading, role in establishing an aquaculture- and fishery-oriented ICM program.

Coordinating and Executing Mechanisms

The main functions of the coordinating body are to promote interagency dialogues on common environmental management issues and to define and coordinate overlapping responsibilities. It also enables the key stakeholders to iron out use conflicts, clarify environmental legislation, and facilitate the execution of multisectoral activities. Thus, it is essential to include the concerned fishery and aquaculture agencies in such a coordinating body.

The action plans should be supported by an executing office that oversees the implementation of the decisions, normally undertaken by the respective line agencies. The specific roles of the coordinating office are enumerated below:

Functions of the Coordinating Office

- coordinates all coastal development activities
- ensures minimization of use conflicts
- prevents and mitigates environmental degradation resulting from economic developments
- approves the strategic environmental management plans and the 5-year action plan
- promotes interagency consultation and the stakeholders' participation
- promotes public awareness and environmental concerns

Aquaculture Zone

One effective way of minimizing use conflicts and protecting the life-support systems for aquaculture is to establish the functional zonation scheme that delineates by legislative measures the areas allocated for development, including the implementation of aquaculture-friendly activities. The setting up of aquacultural zones also provides opportunities for the government and private sectors to systematically plan and develop the allocated areas. In areas where extensive aquaculture operations already exist, the scheme enables the industry to undertake remedial measures necessary to improve farm management in order to reduce the environmental impacts and ensure sustainability. Fish farmers can group together to construct properly engineered systems or form cooperatives to establish proven designs. The scheme will also encourage them to share responsibilities in supervising common water resources, improving marketing and delivery systems, and playing a more significant role in the management of the aquacultural zone as a whole. For example, the shrimp farmers in Surat Thani, Thailand, got together in the 1990s to coordinate the intake and drainage of water to reduce the potential for mixing the two. They used a green flag to indicate when they were taking water into their farms and a red flag to show draining the water from their farms.

At the national level, the existing and potential areas for aquacultural development should be properly mapped out and preferably be included in the plans. However, at this stage, it is difficult to determine the actual boundary of each zone because of the size and length of the coastlines. But at the local level, aquacultural zones should be clearly delineated from areas designated for other users. The designation of aquacultural zones is usually done through the expert advice of aquaculturists and aided by modern tools, such as remote sensing and geographic information system. Establishing the functional zones for aquaculture is a very difficult and time-consuming exercise because of the long process of multisectoral consultation, negotiation, and legalization.

Carrying Capacity of the Culture Area

The environment suitable for aquaculture has a finite capacity for development. The types and intensity of farming operations depend on the environment's capacity either to supply the natural feeds (e.g., nutrient and particulate matters for mollusks and seaweeds) and natural seeds (e.g., fry and fingerlings), or to flush waste material away from the culture sites (e.g., cage and shrimp culture) (Phillips, 1995).

The capacity of the designated aquatic environment to sustain its developmental impact, in addition to other pressures exerted by other human activities, should be carefully measured. However, it is extremely difficult to compute the carrying capacity because of the lack of appropriate methodology to measure accurately the combined environmental absorption capability of an ecosystem or a given water body. At present, the environmental absorption capacity may be measured using certain chemical or biological indicators, including the level of oxygen content in the water, the growth rate of the culture organism, and/or the level of primary production. For example, tracing the growth rate of an indicator species in a culture site will help determine whether the area concerned has reached its maximum carrying limit. On the other hand, a consistent decrease in the size and weight of shrimp in a culture site definitely indicates the lack of sufficient food in the area to sustain the biomass therein.

Knowledge of the carrying capacity is essential in determining the types and sizes of aquaculture operations that could be sustained.

AQUACULTURAL DEVELOPMENT AND MANAGEMENT PROGRAMS

The concerned line agencies of the local governments must take the lead role in developing and managing aquaculture. Adequate development planning within the aquacultural zone is essential, taking into account the environmental guidelines set up by the ICM coordinating body. The guidelines include maintenance of the environmental quality for aquaculture operation, minimization of untreated discharges from shrimp and fishponds in intensive farming systems, and siting and regulation of the density of fish farms and their layout, in accordance with the carrying capacity of the area. The planning process takes into consideration the possible impacts of aquaculture on the environment, as well as the formulation of preventive and mitigating measures against possible adverse impacts of other activities within or outside the zone to the industry's potential development.

Managing the aquacultural zone requires the full cooperation of the fish growers, the other related industries, and agencies of the local government. For a newly designated zone with little or no aquacultural practices, it is rather easy to undertake a comprehensive planning and development program. However, in zones with existing fish farms requiring considerable structural improvements to comply with the environmental quality requirements, the task becomes more difficult and complicated. For example, irregular-sized and poorly constructed fishponds may need major structural improvements. Overcrowded cage farms may be asked to reduce the number of cages or limit new entries to farming. And intensive shrimp ponds may require treatment of their wastewaters before discharge into the open water.

There will be a stricter control of solid waste disposal, and open dumping will not be allowed. All of these actions are constructive measures to reduce pollution and promote sustainability. However, they also mean additional costs to the concerned fish growers. Hence, resistance may develop, rendering the enforcement of such requirements difficult.

It is indeed a challenge to the concerned agencies, and a longer transitional time may be needed to allow the above changes to take place. Although strict enforcement of the regulations is needed for compliance, sufficient technical assistance and economic incentives should also be provided by the government. The change process may be accelerated if fish growers understand the long-term economic benefits they would get by improving the structural design and operation of their farms and the need to maintain their life-support systems.

However, the local governments in most developing nations lack the necessary technical skills and management capability. Thus, a considerable upgrading of institutional capacity at the local level is needed.

Integrated Environmental Impact Assessment

The ICM planning and program development processes normally generate the needed information to assess the cumulative environmental impacts brought about by existing and potential activities in the coastal areas. They also provide a more comprehensive assessment on the state of resources and predictions of future status. This approach, called the Integrated Environmental Impact Assessment (IEIA) differs from the conventional Environmental Impact Assessment (EIA), which normally assesses only the impacts of a single economic activity, such as the EIA of a future cement plant or an oil refinery. The integrated approach, however, considers the absorption capacity of the whole coastal resource system including all the various economic development activities.

Impact assessment should form an integral part of the aquacultural development program within the zone. Its objectives are to

a. determine the changes in the resources systems, in terms of goods and services generated by them
b. assess the level of harm that may befall the habitats the living resources and the human consumers
c. evaluate their socioeconomic implications and recommend preventive or remedial options

The process helps to generate more information and identifies other economic opportunities, in terms of potential areas for coastal tourism, rare wildlife species, untapped water resources, and so on. More importantly, it facilitates the identification and classification of those economic activities that are more harmful to the environment and those that are environment-friendly. In addition, it points out those enterprises that may not be sustainable because of the environmental or socioeconomic problems they may encounter. And finally, impact assessment helps to determine the potential risk to public health coming from development technologies, resource exploitation, and processing.

Impact assessment of the aquacultural zone should be done during the development of the plan. It should take into consideration the environmental impacts of the aquacultural development, as well as the effects of other economic activities on the industry. An overall environmental risk assessment provides information that will help determine the optimal level of development. This provides an opportunity for the industry to select the most appropriate types of aquacultural practices and to determine the level of investments required. However, this does not preclude the requirements for a large aquaculture enterprise to submit an EIA as a condition for awarding the permit.

Environmental Conditions for Aquacultural Development

Water is the life-support system of any aquatic organisms under culture. Thus, ensuring the water quality of the culture site is the fundamental management requirement and role of industry. The aquaculture enterprises therein will eventually collapse if the water quality is allowed to deteriorate, either through self-pollution arising from uncontrolled culture practices or through pollution from external sources.

One way to protect and maintain the environmental quality for aquacultural development is to reduce the damaging impacts of pollution arising from external sources through appropriate siting of industries and controlling and regulating the discharge of domestic and industrial wastes. Equally important is the control of self-pollution arising from waste from aquaculture farms, such as wastewaters from shrimp ponds and fishponds, uneaten feeds from cage farms, and the large amounts of antibiotics and chemicals discharged into the water body.

The introduction of exotic species may also bring about diseases or affect the environmental conditions of the culture sites. Any introduction should be in strict compliance with the international codes of practice (see also Chapter 4).

Aquacultural Products for Human Consumption and Exports

Another threat to sustainable aquaculture is the contamination of products that are harmful to the health of the consumers. Consumption of contaminated products may cause serious illnesses including diarrhea, dysentery, gastroenteritis, hepatitis, cholera, and typhoid. In several countries, many aquacultural products—particularly oysters, mussels, and other mollusks—have been found to be contaminated with high bacteria counts. There are many past instances where live aquacultural products have been banned from import into the United States and Europe. There are also cases where shell production from contaminated sites has been banned completely from entries to the markets of importing nations.

There is a moral responsibility among aquaculturists to ensure that their products are always fit for human consumption. Thus, a stricter control over the production system is needed. This may be facilitated by a comprehensive water-quality monitoring program and the strict imposition of compliance regulation on discharges into the water bodies where aquaculture is practiced.

Water-Quality Monitoring

A basic management requirement is to ensure a healthy culture environment for existing and future aquacultural practices in the designated zones. These include the prevention of pollution arising from external sources and the control of those arising from within the industry. Regular and long-term monitoring of environmental quality changes is essential, especially in areas where there is extensive fish farming.

Regular environmental monitoring enables the early detection of changes in water quality, permits timely management of interventions, and thus avoids costly mitigative measures wherever possible.

The federal government is normally expected to undertake the role of environmental monitoring and overall management of the aquaculture industry. However, because of the lack of human and financial resources, many countries have yet to initiate such a comprehensive monitoring program. The aquaculture industry may be able to play a more distinct role in this direction. (For detail, see Chapter 4.)

Regulatory Requirements and Legislative Responses

Regulatory controls are necessary measures to maintain the harmonious use of natural resources for sustainable development within and outside the industry. For a sound aquacultural development, diverse legal responses to various activities should be coordinated through a specialized legislative framework in order to

a. provide an opportunity for the development of the industry within the designated zones
b. protect and preserve the environmental quality of the designated zones
c. control the carrying capacity of the designated zones in terms of types of species, culture systems, and level of operation
d. provide economic incentives and disincentives for monitoring compliance or efforts to reduce self-pollution
e. control the introduction of exotic species
f. establish the financial mechanism for environmental management and development of the industry

The designated aquacultural zone should be announced by the government as part of its land- and water-use plan. This will provide it with the legal status investors need in order to plan, develop, and manage aquacultural practices in these designated areas. The zoning legislation demonstrates the government's commitment and development priority. More importantly, it creates confidence among the investors because of the importance accorded to the industry. The investors can then expect government support and intervention to ensure that the designated zone will be used exclusively or primarily for aquaculture enterprises.

Legislative requirements are also needed to regulate pollutant discharges from land- and vessel-based activities, as well as from self-pollution. Such requirements may already exist in general terms in the marine environment legislation that includes quality and emission standards (Van Houtte, 1995). They can be used or further refined in order to more effectively protect the designated aquacultural zone from the adverse impacts of industrial, agricultural, and sewage effluents.

National legislation should consider the international dimension of the industry, in terms of the movements of aquacultural produce and species for culture. The international harmonization of the practice is desirable both to secure the quality of aquacultural products and to protect the industry and the environment on which it depends (Howarth, 1995). On the other hand, national legislation should also consider the various national laws and local ordinances governing the diverse and multiple uses of the resources and their economic environment. The integration and harmonization of local and national laws are necessary for any aquaculture legislation to be effective. Such legislation could equally extend to law enforcement. In Xiamen municipality, China, for instance, an integrated law enforcement task team was established only recently to enforce the laws with diverse implications for other economic sectors. Such an integrated team reduces institutional conflicts between various agencies and is an effective mode for law-enforcing operations.

Permit systems are used generally worldwide to regulate the establishment, distribution, siting, size, and operation of the farming entities. This is done most effectively through licensing; as stated by Howarth (1995), the facility to grant or decline license applications provides a great deal of scope to prevent the establishment of undesirable aquacultural projects.

The system helps to confine the aquacultural operations within the limit of the carrying capacity of the designated zone.

It may also be used to provide economic incentives or disincentives to encourage or discourage the development of particular culture practices or species in a particular area. Finally, it is an effective tool for encouraging compliance to pollution-control measures and imposing penalties for noncompliance.

A licensing system has been adopted by many countries in an attempt to regulate the development of the aquaculture industry. In many countries, however, the long and tedious procedure of making multiple applications to regulatory bodies has become an obstacle to potential enterprises. Under the ICM framework, such a problem may be solved through integrated administration of a licensing mechanism that coordinates the roles of other agencies.

However, excessive regulatory controls and enforcement inefficiencies may also restrict the industry's healthy development. A periodic review of existing regulations governing aquacultural development and environmental management is necessary, and should be based on past experiences, changing conditions, and new information that may help improve their efficiency and effectiveness.

Information Base for Aquacultural Development and Management

Aquaculture operations are highly localized, and their management must be in harmony with local ecological, sociocultural, and economic conditions. Building a strong database on the local conditions should be an integral part of the government's aquacultural development and management program. The essential database needs to include information on the farming systems (i.e., distribution, size, culture systems, species, and operation), financial investment (i.e., level of investment, ownership, size, and source of funds), operators (i.e., skilled or unskilled, number, gender, age, and income level), products market (i.e., delivery channel, postharvest, and export); socioeconomic benefits (i.e., source of food, employment, foreign-exchange earnings, and distribution), and constraints (i.e., losses due to natural hazards, diseases, use conflicts, political or cultural restrictions, pollution, and so on). This is the information that must be gathered, updated, and analyzed for timely policy and management interventions.

Lack of good databases often places the aquaculture industry in a weak position in terms of national and local governments' funding allocation and support, especially when competing with other sectors for limited natural and financial resources. Hence there is a need to acquire a reliable information base and the ability to use it wisely for the management of the industry, as well as for ensuring its socioeconomic importance in the rural economy.

National and local fishery and aquaculture programs should be reoriented to include research that responds to management needs, especially those focusing on resolving developmental and environmental problems. The report on a recent technical workshop (FAO and NACA, 1995) concluded that precise and quantified information on the interactions between aquaculture and the environment was not available. The report pointed out that, while the nature of the impacts was generally known, the knowledge of what caused the impact was often unquantified and speculative. There was also a lack of precise and quantified information on the environmental impacts caused by aquacultural development. Research into such problems would certainly help enhance national policy and farm-management measures. In addition, the workshop prepared a detailed list of research sources that support policy development.

There is also a need to narrow the gap in aquacultural practices between research and the farm sectors so that the results of studies can be readily utilized. As cited by Fegan (1995), research studies have shown that bacteria isolated from shrimp and water samples were resistant to oxytetracycline. Despite this finding, the antibiotic is still widely used by many shrimp farmers. Efforts must be made to simplify the research results so that they can be easily understood and applied by aquaculture farmers. The important role that industry plays in aquacultural research should be given due recognition by the government; likewise, appropriate support from the public sector is necessary to encourage the sharing of research results.

Capacity Building at the Local Level

One major constraint to the industry's sustainability is the lack of a critical mass of aquacultural and coastal zone managers at the local level who can plan and implement the activities of a sound development and management program (Crawford et. al., 1993; Schroder, 1993). Existing university training has not been able to provide this pool of experts. The lack of expertise at the university teaching-staff level makes it difficult for the traditional training institutions to meet the above needs. Specialized trainings need to be established with a broad-based management curriculum that caters especially to the development of management skills in coastal planning and management, including those for aquaculture.

A budgetary allocation to develop a critical number of a multidisciplinary management team members at the local level is a wise public investment, considering the substantive ecological and socioeconomic benefits to be gained from a skillful management of coastal resources.

Public Awareness

The aquaculture industry may become sustainable only if there is a more equitable sharing of socioeconomic benefits among the various resource users. However, the functional integrity of ecosystems should not be sacrificed, and there should be minimal conflicts among competing resource users. To achieve this goal, public endorsement and the support of the aquaculture industry are essential.

Special efforts should be made to enhance environmental awareness among aquaculturists and the manufacturers of feeds, chemicals, and drugs. All of them must closely cooperate with farmers through on-farm research and other participatory activities.

Both the investors in the industry and the small-scale fish farmers need to be informed of government plans on aquacultural development. Whenever possible, both groups can contribute to the planning and development of major aquaculture projects or programs. Equally important is the public at large, who should know what the industry is generating and what they can expect from it. A public-awareness campaign is an important tool for soliciting popular support for healthy products by protecting and ensuring good water quality for aquaculture operations. It can also be a powerful tool to protect aquacultural products from unhealthy culture conditions. Thus public-awareness campaigns must form a substantive part of any coastal or aquacultural management program.

Financing the Management of Aquacultural Zones

A fundamental change of approach must be developed at the governmental and industry levels, if aquaculture is to continue playing an important role in national economies. A financial mechanism should be established by the government in collaboration with the industry to pool the necessary financial resources for maintaining a culture environment that is conducive to existing and future expansion of aquaculture in designated zones. The cost of providing such services should be included as part of the taxes collected from the industry. Thus, an overall review of the current taxation scheme for aquacultural practices is necessary, especially in view of the fact that a large number of existing coastal practices are made up of small farms.

Despite aquaculture's substantive contribution, especially to export earnings, only a small part of the revenues generated from it has been used to ensure the healthy growth of the industry. Unless a healthy environment is maintained, this industry will not be able to sustain its current growth. Equally important is for the aquaculture industry to take the necessary initiatives to collectively protect the common culture environment on which its very existence depends.

Management of the aquacultural zones requires technical inputs, in terms of personnel and facilities, in order to monitor environmental quality, enforce regulations, and respond to disease outbreaks and natural hazards. Thus, there must be sufficient financial means to provide the needed services, sustain existing aquacultural practices, and develop new ones. Sustainable financing mechanisms must be developed and provided with the necessary legal support for their operations. Financing the environmental management of the aquacultural zones should be the combined responsibility of the government and the industry. The sources for financial supports other than direct taxes need to be explored, including water-use fees, pollution fees, permit or licensing fees, contributions, and donations.

CONCLUSIONS

Integrated management is not a new concept, but ICM programs are difficult to develop and implement because many users compete over the finite marine resources. The inclusion of marine resources in local economic development planning in the coastal region is indeed a new approach, arising from awareness of the socioeconomic significance of maritime industries and realization of the existence of the rich but fragile ecosystems of the coastal zones. There are increasing efforts to include sea-use planning in national development programs.

More than 100 ICM programs have already been initiated in the last two decades in many parts of the world (Sorensen, 1993); still, there is much to be learned from their various successes and failures about developing and implementing ICM programs. ICM has evolved into a system whose application may help harmonize the competing users of natural resources, promote sustainable development at the local level, reduce social conflicts, and protect the functional integrity of the important ecosystems that generate goods and services for humankind.

Experience has shown that ICM is most effective at the lowest level of the local government's administration. For example, an ICM program for ecosystem management has proven to be very effective if the community can be mobilized to manage the natural resources on which their livelihood depends. The management of Apo Island in the Philippines was successful because almost all of the 200 or more resident families were directly or indirectly dependent on the coral resources for their livelihood (White, 1987).

Application of the ICM system for coastal aquacultural development and management can also be effective in protecting the interests of the aquaculture sector in a coastal area where other users are competing for space and resources. Proper planning and management enable the designation of coastal and marine resources for specific uses and thus provide an opportunity for the concerned sector to develop.

The development and sustainability of aquaculture largely depend on the ability of the industry to neutralize the adverse impacts of externalities (Wijkstrom, 1995). ICM is the right tool for the allocation of resources; for integrated planning and control of resource use; and for coordination and integration of policy, management, and technological instruments.

Two factors have made coastal aquaculture an increasingly important industry. The first is the demand for an increasing fish supply to augment the limited world production resulting from capture fisheries. The second factor, which relates to the first, is the increase in fish prices, which will make aquacultural practices profitable enterprises. However, coastal aquaculture also faces two significant constraints, namely: the increasing contamination of coastal waters, and the greater competition for marine resources by other sectors. There is no trend toward improvement of coastal environmental quality, especially in the Asian region (Gomez, et al., 1990, ESCAP and ADB, 1995), which is limiting the available spaces and areas for fish cultivation. In many countries, areas suitable for aquaculture are also heavily utilized for other economic activities, thus increasing conflicts over resource use.

ICM provides a solution to these problems through an integrated planning and management of resource use; but, more importantly, it must be backed up by strong policy and institutional arrangements in order to enable the application of appropriare management interventions. Despite the difficulties encountered in the application of ICM for the sustainable development of coastal aquaculture, the pressures for increased fish supply, employment, and foreign-exchange earnings support the development and maintenance of the industry. However, we should realize that the unique life-support system on which this industry heavily depends cannot yet be replaced. Thus, the fundamental step toward ensuring a sustainable aquaculture industry is to prevent the continuous degradation of marine environmental quality.

The future of the coastal aquaculture is bright, but it greatly depends on how we treat our marine environment.

REFERENCES

ADB (Asia Development Bank) and NACA (Network of Aquaculture Centres in Asia). 1991. Fish health management in Asia Pacific. Report on a Regional Study and Workshop on Fish Diseases and Fish Health Management. ADB Agriculture Development Report Series 1. Network of Aquaculture Centres in Asia Pacific, Bangkok, Thailand. 627 p.

Boromthanarat, S. 1995. Coastal zone management. pp. 431–434, In: FAO/NACA. Report on a Regional Study and Workshops on the Environmental Assessment and Management of Aquaculture Development. Food and Agriculture Organization of the United Nations, Bangkok.

Chien, Y. H. 1992. Aquaculture in The Republic of China: a biosocioeconomic analysis of the aquaculture industry in Taiwan. pp. 31–50, In: I-Chiu Liao, Chung-Zen Shyu, and Nai-Hsien Chao (Eds.). Aquaculture in Asia. Proceedings of the 1990 APO Symposium in Aquaculture. September 5–13, 1990, Taiwan Fisheries Research Institute, Taipei.

Chua, T.-E. 1986. An overview of the fisheries and aquaculture industries in Asia. pp. 1–8, In: J. L. Maclean, L. B. Dizon, and L. V. Hosillos (Eds.). First Asian Fisheries Forum. Asian Fisheries Society, Manila, Philippines.

Chua, T.-E. 1990. Environmental management of coastal aquaculture development. Conference on Environmental Issues in the Third World Aquaculture Development, September 17–22, Bellagio, Italy.

Chua, T. E. 1992. Coastal aquaculture development and the environment: the role of coastal area management. Mar. Pollut. Bull., 25(1–4): 98–103.

Chua, T.-E. 1993. Elements of integrated coastal zone management. Ocean and Coastal Manage., 21: 81–108.

Chua, T.-E. 1994. Asian fisheries towards the year 2000: a challenge to fisheries scientists. pp. 1–14, In: L. M. Chou, A. D. Munro, T. J. Lam, T. W. Chen, L. K. K. Cheong, H. W. Khoo, V. L. Phang, K. F. Shim, and C. H. Tan (Eds.). Third Asian Fisheries Forum. Asian Fisheries Society, Manila, Philippines.

Chua, T.-E., and L. R. Garces. 1994. Marine Living resources management in the Asean Region: lessons learned and the integrated management approach. Hydrobiologia, 285: 257–270.

Chua, T.-E., and J. N. Paw. 1987. Aquaculture development and coastal zone management in Southeast Asia: conflicts and complementarity. pp. 2007–2021, In: O. T. Magoon, H. Converse, D. Miner, L. T. Tobin, D. Clark, and G. Domurat (Eds.). Proceedings of the Fifth Symposium on Coastal and Ocean Management, Vol. 2. American Society of Civil Engineers, New York.

Chua, T. E., J. N. Paw, and F. Y. Guarin. 1989. The environmental impact of aquaculture and the effects of pollution on coastal aquaculture development in Southeast Asia. Mar. Pollut. Bull., 20(7):335–343.

Crawford, B. R., J. S. Cobb, and Friedman, A. 1993. Building capacity for integrated coastal management in developing countries. Ocean Coastal Manage., 21: 311–338.

ESCAP (Economic and Social Commission for Asia and the Pacific) and ADB (Asia Development Bank). 1995. State of the marine environment in Asia and the Pacific. United Nations, New York. 651 p.

FAO (Food and Agriculture Organization of the United Nations). 1991. Living marine resources. Research paper 5 commissioned by the United Nations Conference on Environment and Development (UNCED), Rome. 41 p.

FAO and NACA. 1995. Report on a regional study and workshops on the environmental assessment and management of aquaculture development. Food and Agriculture Organization of the United Nations, Bangkok. 492 p.

Fegan, D. F. 1995. Environmental assessment and management of aquaculture development: an industry perspective. pp. 437–445, In: FAO and NACA. Report on a regional study and workshops on the environmental assessment and management of aquaculture development. Food and Agriculture Organization of the United Nations, Bangkok.

GESAMP (IMO, FAO, Unesco, WHO, WMO, IAEA, UN, and UNEP). 1992. Joint group of Experts on the Scientific Aspects of Marine Pollution. Reducing Environmental Impacts of Coastal Aquaculture. Reports and Studies GESAMP 47. 35 p.

Goldberg, E. D. 1994. Coastal zone space—Prelude to conflict? United Nations Educational, Scientific and Cultural Organization (UNESCO), Paris. 138 p.

Gomez, E. D., Deocadiz, E., Hungspreugs, M., Joathy, A. A., Jee, K. K., Soegiarto, A., and Wu, R. S. S. 1990. State of the Marine Environment in the East Asian Seas Region. UNEP Regional Seas Reports Studies 126. 63 p.

Han, B., and Huang, T. 1990. Green oysters caused by pollution on the Taiwan coast. Environ. Pollut., 65: 347–362.

Howarth, W. 1995. The essentials of aquaculture regulation. pp. 459–465, In: FAO and NACA. Report on a regional study and workshops on the environmental assessment and management of aquaculture development. Food and Agriculture Organization of the United Nations, Bangkok.

Kim, In-Bae. 1992. Aquaculture in the Republic of Korea. In I-Chiu Liao, Chung-Zen Shyu, and Nai-Hsien Chao (Eds.). Aquaculture in Asia: Proceedings of the 1990 APO Symposium in Aquaculture. September 5–13, 1990, Taiwan Fisheries Research Institute, Taipei.

Liao, I. C. 1990. Aquaculure in Asia: status, constraints, strategies, and prospects. pp. 13–27, In: I-Chiu Liao, Chung-Zen Shyu, Nai-Hsien Chao (Eds.). Aquaculture in Asia: Proceedings of the APO Symposium in Aquaculture. September 5–13, 1990, Taiwan Fisheries Research Institute, Taipei.

Miki, K., and M. Sano. 1992. Aquaculture in Japan. pp. 81–89, In: I-Chiu Liao, Chung-Zen Shyu, and Nai-Hsien Chao (Eds.). Aquaculture in Asia: Proceedings of the 1990 APO Symposium in Aquaculture. September 5–13, 1990, Taiwan Fisheries Research Institute, Taipei.

NEDA (National Economic Development Authority) 1992. The Lingayan Gulf Coastal Area Management Plan National Economic Development Authority, Region 1, Philippines. ICLARM Technical Report 32. 87 p.

Pauly, D., and T.-E. Chua. 1988. The overfishing of Southeast Asian marine resources: socioeconomic background in Southeast Asia. Ambio, 19(3): 200–206.

Paw, J. N., and T.-E. Chua. 1991. An assessment of the ecological and economic impact of managrove conversion in Southeast Asia. pp. 201–212, In: L. M. Chow, T.-E. Chua, H. W. Khoo, P. E. Lim, J. N. Paw, G. T. Silvestre, M. J. Valencia, A. T. White, and P. K. Wong (Eds.). Towards an integrated management of tropical coastal resources. ICLARM Conference Proceedings 22. National University of Singapore, Singapore National Science and Technology Board, Singapore; and International Center for Living Aquatic Resources Management, Philippines. 455 p.

Phillips, M. J. 1995. Environmental aspects: regional overview. pp. 423–429, In: FAO and NACA. Report on a regional study and workshops on the environment assessment and management of aquaculture development. Food and Agriculture Organization of the United Nations, Bangkok.

Pido, M. D., and Chua, T.-E. 1992. A framework for rapid appraisal of coastal environments. pp. 135–148, In: Chua and Scura (Eds.). Integrated framework and methods for coastal area management. ICLARM Conference Proceedings 37. Manila.

Scura, L. F., T.-E. Chua, M. D. Pido, and J. N. Paw. 1992. Lessons for integrated coastal zone management: the Asean experience. pp. 1–69, In: T.-E. Chua and L. F. Scura (Eds.). Integrative framework and methods of coastal area management. ICLARM Conference Proceedings 37. Manila.

Schroder, P. C. 1993. The need for international training in coastal management. Ocean Coastal Manage., 21: 303–310.

Sorensen, J. 1993. The international profliferation of integrated coastal zone management efforts. Ocean Coastal Manage., 21: 45–80.

Tokrisna, R. 1992. Aquaculture in Thailand. pp. 113–124, In: I-Chiu Liao, Chung-Zen Shyu, and Nai-Hsien Chao (Eds.). Aquaculture in Asia: Proceedings of the APO Symposium in Aquaculture. Taipei Fisheries Research Institute, Taipei.

Tokrisna, R. 1995. Economics of planning for multiple coastal and marine resources uses. pp. 471–477, In: FAO and NACA. Report on a regional study and workshop on the environmental assessment and management of aquaculture development. Food and Agriculture Organization of the United Nations, Bangkok.

Vallejo, S. M. 1993. The integration of coastal zone management into national development planning. Ocean Coastal Manage., 21: 163–182.

Van Houtte, A. 1995. Fundamental Techniques on Environmental Law and Aquaculture Law. pp. 451–457, In: FAO and NACA. Report on a regonal study and workshops on the environmental assessment and management of aquaculture development. Food and Agriculture Organization of the United Nations, Bangkok.

White, A. T. 1987. Coral reefs: valuable resources of Southeast Asia. ICLARM Education

Series 1. International Center for Living Aquatic Resources Management, Manila, Philippines. 36 p.

Wijkstrom, U. 1995. Environmental management and aquaculture: the issues from the perspective of an economist. pp. 467–470, In: FAO and NACA (Eds.). Report on a regional study and workshops on the environmental assessment and management of aquaculture development. Food and Agriculture Organization of the United Nations, Bangkok.

Wu Bao-Ling. 1991. Pollution has damaged coastal aquaculture. Mar. Pollut. Bull., 22: 371–372.

9 Planning, Regulation, and Administration of Sustainable Aquaculture

JOHN S. CORBIN AND LEONARD G. L. YOUNG

INTRODUCTION

Aquaculture is maturing as a science and business. It strongly influences environmental change in developing nations and alters management approaches to aquatic resources in developed nations. Driving this rapid advance has been widespread experimentation to control the life cycles of economically important aquatic species, and development of standardized commercial protocols and systems to maximize yields and profits. Complementing this explosion in critical knowledge are improved mechanisms to disseminate technical information globally, finance large development projects, and market the resulting fish and shellfish supplies internationally.

The 21st century promises greater growth and diversification, as new technologies become practical and new species become available. Capture fisheries continue to decline, further increasing demand, and national governments are increasingly turning toward aquaculture for food and economic needs. From patterns of world development in recent years, it is clear that increasing production will necessitate the dedication of ever larger amounts of vital natural resources for aquacultural use; these resources include coastal lands, groundwaters, rivers, lakes, and nearshore marine habitats, and therefore, governmental oversight.

Recent experience indicates that unbridled aquacultural development can cause serious environmental degradation as well as have unanticipated economic, social, and cultural impacts on surrounding communities. Mounting evidence demonstrates that poorly planned, sited, and managed aquaculture enterprises, while achieving short-term economic goals, are not sustainable for the long term (Pullin et al., 1993; FAO and NACA, 1995). If aquaculture is to fill the emerging gap in world aquatic protein supply, governments must begin to plan and manage a plethora of culture species, systems, and environments for sustainable production over a long time

Sustainable Aquaculture, Edited by John E. Bardach
ISBN 0-471-14829-6

period. The urgency to develop workable, cost-effective, widely applicable approaches to sustainability is particularly critical for Asia, where 80% of future aquaculture expansion is projected to occur (Nash, 1987).

Government's central and critical role in guiding, directing, and monitoring the aquaculture industry's resource utilization is important to achieving expansion in a way most beneficial for society. This chapter describes contemporary assessment frameworks, administrative systems, and new conceptual approaches to planning, developing, and managing sustainable use of the earth's finite natural resources for aquaculture. Aquaculture-related examples are described when available. Information from both developed and developing countries is included for the broadest applicability.

SUSTAINABILITY AS THE GOAL

Government's Role

In many nations, sustainable development has become the long-range goal for managing how humans relate to their fragile and finite natural environment. Governments today are urgently examining their roles in achieving sustainable, long-term use of resources, as well as in identifying desirable industries that meet society's economic expectations and quality-of-life requirements. With increasing global population, expanding the supplies of aquatic protein through aquaculture is a critical need, but it should not be done at the expense of the environment.

The goals of sustainable development and aquacultural expansion provide an understanding of how existing aquacultural farms around the world should perform and how new farm expansion should be planned and managed. Sustainable aquaculture should

- conserve natural resources and biodiversity
- achieve the least degradation of the environment
- utilize techniques and technologies appropriate to a situation and site
- generate profit or economic benefits in excess of costs
- foster minimal social disruptions and conflicts
- provide for community needs

The public sector, with its broad empowerment to act for the common good, plays a central role in setting the national agenda for aquaculture and its sustainable development. A national government that targets sustainability should utilize its power, influence, and resources to assure that industry expansion embraces the described characteristics through direct and indirect intervention in project planning, design, and implementation.

TABLE 9.1. Comparison of Commonly Observed Elements of Long-Term Aquacultural Vision in Developing and Developed Countries

Developing Country

- Rural economic development, providing alternative employment opportunities
- Better nutrition for rural poor
- Increased export earnings
- Attraction of foreign investment
- Recycling of agricultural and domestic wastes
- Integrated agriculture-aquaculture systems
- Increase in domestic food supplies of popular local species

Developed Country

- Rural economic development, replacing agricultural jobs
- Increased variety of seafood supplies
- Import substitution
- Trade imbalance in seafood addressed
- Use of recirculating and water-conservation culture systems
- Development of quality assurance programs
- Recreational fisheries enhancement
- Commercial fisheries enhancement

Comparing Visions for Aquacultural Development

The policy structure, statutory and regulatory framework, and public support programs implementing development efforts form a country's broad national vision for aquacultural expansion. Historically, it is important to note that developing and developed countries have had somewhat different long-term visions for aquacultural farming and its expected contribution to society (Table 9.1).

Many developing nations are troubled by the strong trend of rural populations to migrate to overcrowded cities because of insufficient opportunities to earn money to support large families (Postel, 1996). Economic development and the creation of appropriate employment opportunities in rural areas have been central objectives for aquaculture in developing countries. Regional settlement issues—that is, balancing the population distribution with more equitable sharing of the national wealth—will continue as important reasons to encourage further expansion.

Developing-country governments have strongly encouraged the aquaculture of products that can be exported in order to generate important foreign exchange for importing goods and buying services to supply other national needs. Complementing this objective is the desire to attract foreign investment capital from private sources and development banks (e.g., the Asian Development Bank) to help finance large-scale developments. In combination, these two directives have tended to focus species choice on products (e.g., marine shrimp) consumed by the industrialized countries.

For Africa, in contrast, rural development assistance has focused on improving basic protein supplies and the nutrition of numerous, isolated inland villages. Subsistence aquaculture (e.g., low-input, small-scale production systems) is encouraged as another component of a diverse forest-gathering, multilivestock, multicrop, rural family food production system. In this setting, a rural farmer views small-scale aquaculture as a way of spreading the ever-present risk of starvation (UNDP, NMDC, and FAO, 1987). Low-input production technologies for local species in contrast to high-yield, export-oriented farming, entail using indigenous materials, recycling agricultural and domestic wastes, and integrating agriculture-aquaculture systems.

Aquaculture in developed countries has a strong economic focus, with an entrepreneurial, for-profit business emphasis. The basic availability of food and adequate community nutrition are not critical issues. Instead, government planning has stressed the desire to increase the quantity and variety of seafood available to consumers, while substituting for foreign imports and exporting to increase international trade.

Providing rural agriculture and fishing communities with alternative means for producing income and maintaining lifestyles, is a growing concern in the United States, Norway, Scotland and other developed countries. These small, isolated communities of relatively independent farmers and fishers are being forced out of business by fast-changing technologies, economics, and markets. Concerned governments, with the support of the farmers and fishers, wish to replace the existing job and tax base with compatible, economically viable industries (Pillay, 1990).

Several economically important and growing dimensions of developed-country aquaculture serve large segments of society, though they do not directly produce and market food. The culture of freshwater game fish in public and private hatcheries, with their subsequent release into lakes, rivers, and streams, supports recreational fishing in many developed countries. For example, in the United States over a billion hatchery fish are released annually, and 60 million recreational anglers generate over $30 billion a year in revenue (AFS, 1990). Rapid advances in culture technologies for saltwater species are adding new opportunities to those successes clearly demonstrated by many salmon stock-enhancement programs that serve both recreational and commercial harvesting purposes. Moreover, in a few tropical areas of the world (e.g., in Hawaii), experimental culture of stony corals and reef fishes points to large-scale restoration projects for tropical habitats supported by aquaculture (Corbin and Young, 1995).

SITE ASSESSMENT FOR NEW FACILITIES

The Challenges of Sustainability and the Problems to Resolve

Review of recent world developments provides numerous examples of detrimental environmental, economic, and sociocultural impacts attributed to poorly planned

TABLE 9.2. Major Issues and Problems in Sustainability of Contemporary Aquaculture in Developed and Developing Countries

Category	Issue or Problem	Country	
		Developed	Developing
Biophysical	Overconversion of mangroves		X
	Overconversion of wetlands		X
	Water quality and substrate degradation	X	X
	Overpumping of groundwater and saltwater intrusion	X	X
	Toxic algal blooms	X	X
	Disease contamination from cultured stocks	X	X
	Reduction of wild catches	X	X
	Genetic dilution of native stocks	X	X
	Exotic species introduction	X	X
	Misuse of antibiotics		X
	Wildlife impacts	X	
Economic	Multiple use of resources conflicts	X	X
	Displacement of traditional subsistence and cash activities	X	X
	Diversion of inputs (e.g., feed ingredients) into aquaculture		X
	Unsuitable enterprise development models		X
	Focus on exportable species		X
Sociocultural	Competition for resources		X
	Visual pollution	X	
	Acceptance of innovation and technology		X
	Altered familial work patterns		X
	Increased under- and unemployment		X
	Degradation of community nutrition		X
	Religious and cultured taboos		X

and managed aquaculture (Pillay, 1992; Pullin et al., 1993; FAO and NACA, 1995). Highlighting these issues provides comparison of the sustainability goals of developed and developing countries, as well as a partial list of real problems, which may be addressed through governmental action (Table 9.2).

Biophysical Problems

Biophysical problems are the most prevalent and widely recognized results of unsustainable aquacultural development (Pillay, 1992; Austin, 1993; Phillips, 1993; Pullin, 1993; Rosenthal, 1994; FAO and NACA, 1995). The details are given in Chapters 1 and 4.

An expanding issue, particularly in the United States and Europe, is farm damage and stock loss due to the feeding of legally protected wildlife, such as seals and otters on salmon net pens and herons and cormorants on pond-reared animals. The inability of farmers to adequately manage predation has caused serious economic loss and forced legal confrontations with wildlife officials (Pillay, 1992).

Economic Problems. The economies of small rural communities may be severely disrupted by large-scale commercial aquaculture or the externalities related to its development. Competition for space, resources, and labor has resulted in serious conflicts in both developing and developed countries. Multiple-use conflicts stem from competition for the same water resource or space, which historically has been uncontrolled and considered common property. Often these complex issues revolve around the desired collocation of aquaculture with incompatible activities, for example

- industrial use of waterways to transport goods or discharge wastes
- agricultural uses that draw water to irrigate food crops
- recreational users who desire unimpeded boating or fishing
- coastal-dependent fishery or tourism uses (Ackefors and Grip, 1995).

Conflict resolution and direct intervention by the government are often necessary to optimize multiple-use situations and resolve user problems.

Development of export-oriented aquaculture in rural, developing communities has caused disruption of traditional systems of commerce, which depend on a range of subsistence and low-income activities (Weeks, 1990; Landesman, 1994). Problems may result from applying the Western business development model of full-time employment to developing, rural communities, where a part-time or seasonal, microenterprise approach to aquacultural development is more appropriate (Newkirk, 1993). Large-scale aquacultural farming, small-scale owner-operated farms, and subsistence aquaculture ventures are qualitatively very different in their community interactions, with economic impacts on, and acceptance by rural societies (Weeks, 1990). Furthermore, focusing production on exportable species (e.g., marine shrimp) for foreign exchange may also divert investment capital and resources from the culture of less-valuable local species, thus limiting the availability of domestic supplies. For further details, see Chapters 2 and 6.

Sociocultural Problems. A community's inherent social system and cultural beliefs can strongly influence aquaculture project success, be it a large-scale commercial venture or a small-scale community-based activity (Pollnac et al., 1982). When aquaculture competes favorably with agriculture, commercial fishing, mangrove harvesting and other activities, access to land and water resources by the rural poor is often restricted (Shang, 1990). In the developed world, differing perceptions of the aesthetics of nearshore cage and pen farming systems have caused heated siting controversies in coastal areas of North America and Europe (Gowen and Rosenthal, 1993).

Where aquaculture has not been traditionally practiced—as in parts of Africa—it is a relatively new technology and can prove difficult to transfer to rural peoples.

Development proponents have lacked understanding of a community's social and family structure and appreciation of indigenous "technical" knowledge, both leading to project failures (Weeks, 1990; Ruddle, 1993). Patterns of rural employment and nutrition can also be negatively changed such that local residents suffer under- or unemployment when traditional activities are replaced by low-wage employment on aquacultural farms or processing plants (Landesman, 1994). Nutritional patterns may also be altered, as women who prepare food must now carry out more of their husband's duties on the family farms (Weeks, 1990). Successful aquacultural development in a region can also be greatly affected by local religious and cultural beliefs. For example, rearing fish for consumption is still considered to be against the basic principles of Buddhists and Hindus, who are vegetarians (Shang, 1990); obviously, ways around this proscription have been found. For further details, see Chapter 2.

Assessment Approaches

Macrolevel: Country and Regional Approaches. Use of large-scale environmental assessments for aquaculture is beginning to be supported by a variety of remote-sensing (surveying and mapping) technologies (Table 9.3), and these have vastly expanded our capability to assess and monitor the earth's finite resources (Ikeda and

TABLE 9.3. Selected Resource Assessment Applications of Remote-Sensing Technology

Agriculture:	Environmental management:
Crop inventory	Water-quality assessment and planning
Crop yield prediction	Environmental and pollution analysis
Assessment of flood damage	Coastal-zone management
Soil inventory	Wetlands and mangrove forest mapping
Forestry and rangeland:	Lake-water quality
Forest habitat assessment	Resource inventory
Wildlife range assessment	Water resources:
Land resource management	Surface-water inventory
Land-cover inventory	Flood control and damage assessment
Comprehensive use planning	Monitor runoff and pollution
Facility siting	Water circulation, turbidity, and sediment
Floodplain delineation	Lake eutrophication survey
Fish and wildlife	Groundwater location
Wildlife habitat inventory	Ocean Resources
Wetlands location, monitoring, and analysis	Sea temperature
Vegetation classification	Surface color and biological productivity
	Wave height
	Surface winds
	Surface topography
	Currents and circulation

Source: OTA (1993a).

Dobson, 1995; Verbyle, 1995; Kew, 1996). As of 1993, five nations (France, Japan, India, Russia, and the United States) had placed environmental monitoring satellites in geosynchronous orbits (OTA, 1993a). A wide variety of satellite information is available directly from the U.S. government or through many private companies (Baker, 1996; Beaujardiere, 1996; EDC DAAC, 1996; Hall, 1996). Low-level remote sensing is a less costly means than satellite monitoring for assessing environmental conditions over large areas. For developing countries, greater accessibility to small planes, special cameras, and film processing is reducing costs and the need for outside expertise to collect and interpret information.

Geographic information systems (GIS) use computer-assisted systems to store and interpret remotely sensed data as well as on-ground and other environmental information (McGuire et al., 1991; Kam et al., 1992). GIS hardware and software system requirements are becoming more powerful, widely available, specialized, and cheaper each year. In recent years, the use of GIS has expanded from land management to applications in the social sciences, aquatic sciences, and economics (Lyon and McCarthy, 1995), all of which are important dimensions of sustainable aquaculture (Table 9.4). The major benefits of using GIS lie in its ability to help in

- handling extremely large data volumes
- integrating various kinds of data (graphic, textual, digital, analog)

TABLE 9.4. Use of GIS for Assessing Potential Aquaculture Sites

Key Area	Database Descriptions
Biophysical	Salinity variations
	Soil characteristics and types
	Rainfall patterns
	Mangrove coverage
	Hydrographic factors
	Coastal geomorphology
	Meteorological characteristics
	Water source and quality
	Critical habitats
Socioeconomic	Existing land uses
	Existing ocean uses
	Population distribution
	Employment statistics
	Income levels
	Educational background
	Infrastructure availability
	Demographics
	Cultural beliefs
	Religion

- greatly enhancing the capacity for data exchange
- processing and analyzing data more efficiently and effectively than by manual procedures
- modeling, testing, and comparing alternative scenarios
- efficiently updating data, especially graphic information

These large-scale and expensive techniques offer government and large corporations powerful tools to plan, monitor, and manage natural resources for both consumptive and nonconsumptive uses, including aquaculture (Zeylmans, 1996). Perhaps the greatest advantage is that various sources of information can be easily integrated and manipulated for timely decisions. However, the availability of skilled personnel is an ongoing problem, especially in developing countries (Conant et al., 1983; Kam et al., 1992).

Microlevel: Local and Project-Level Planning. Planning a specific aquaculture project, much less sustainable one is a complex, interdisciplinary task. Successful aquaculture at any location involves both technological and nontechnological ingredients that must be assessed, planned for, decided on, and managed using a holistic approach (Table 9.5). The most obvious factors among these are the physical, location-dependent elements of land, water, and climatic resources. Other site-dependent, nontechnological determinants, often not fully appreciated, include economics, markets, community sociology and culture, governmental policy, regulation, and politics (Table 9.6). Experience shows that failure in, or a major conflict with, any one of these components can slow or stop project development or cause a successful operation to fail and not be sustainable (Pullin et al., 1993; Landesman, 1994; Aiken and Sinclair, 1995).

TABLE 9.5. Major Species and Technology Selection Factors and Choices for Aquaculture Project Planning

Factors	Choices
	Species Selection
Water temperature	Warm- or cold-water species?
Water type	Freshwater, brackish, or seawater?
Biotechnology	Species maturation, spawning, and nutrition known?
Marketability	Local or export potential known?
	Production Technology Selection
Location	Land-based or nearshore?
Management intensity	Extensive, semi-intensive, or intensive?
Scale	Backyard or subsistence; small, medium, or large?
Integration	Polyculture, multi-culture (growing species, individually, successively in the same water), or integrated agriculture-aquaculture?

TABLE 9.6. Key Site Assessment Factors Affecting Sustainability of Aquaculture Projects in Developing and Developed Countries

Primary Factors	Sustainability Issues
	Biophysical
Climate:	
Sunlight	Growth rates, pond productivity
Rainfall	Pond salinity, flooding
Air temperature	Growth rates
Wind	Water temperature and oxygenation
Land:	
Soil type	Water loss, pumping costs
Soil chemistry	Toxicity, water quality, growth rates
Percolation rate	Water loss, pumping costs
Source water:	
Temperature	Growth rates
Availability	Water quality and productivity
Chemistry	Water quality, growth rates
Disposal	Environmental and farm/self-pollution
	Economic
Infrastructure:	
Seed and feed	Availability and cost
Roads, rail, harbors	Transportation options
Utilities	Technology options, problem resolution
Communications	Technology options, problem resolution
Credit	Farm expansion, problem resolution
Labor:	
Income sources	Competition for labor
Availability and distribution	Adequacy of labor force
Market:	
Demand	Sales and distribution options
Location	Transportation costs
Land and water availability:	
Ownership	Long-term commitment to project
Existing uses	Competition for use
Wetlands and mangrove usage	Ecological and community impacts
	Sociocultural
Institutions, formal and informal	Compatibility and access to resources
Social patterns:	
Benefits distribution	Motivation and commitment
Male/female labor patterns	Community and family support
Societal structure and leadership	Community approval and support
Belief systems	Compatibility
Traditional knowledge, skills, customs, practices, and values	Conflicting knowledge and use
Experience with technological change	Acceptance and long-term use

Hawaii's approach to long-range aquacultural planning utilizes a holistic approach to comprehensively describe specific aquacultural development niche opportunities (Corbin and Young, 1988; ADP, 1993). Other recent applications of this "overarching" concept for natural resource planning and management include integrated coastal-zone management for coastal resources (Chua and Scura, 1992) and agricultural research and extension (DeBoer, 1992). For further details on biophysical and economic factors, see Chapters 2, 4, 6, and 8.

Sociocultural information is more critical to success in rural developing locations, where traditions and customs play a central role in the workings of society and are perhaps the most difficult and least-addressed assessment area. Only recently have social scientists and cultural anthropologists studied the impacts of aquaculture on rural societies, particularly in developing countries. The diversity of human communities and the complexity of behavior, particularly their response to change and innovation, allows only some broad generalizations regarding a few pertinent factors for study.

Experience demonstrates that projects being developed by "outsiders" will need a staff member who knows and is known to the community to properly carry out any assessment. Rural communities typically have formal and informal institutions that deal with property and use rights. These systems determine what type of activities could be undertaken on nearby land and nearshore waters, when, and by whom (Roy, 1985). Outside developers should be aware of traditional systems of resource allocation, and should engage in a dialogue with controlling interests in order to reach common understandings and accommodations. Failure to understand these issues can negatively affect project acceptance by the community.

Deep-rooted social patterns should be carefully studied, particularly for subsistence-level and small-scale aquacultural development initiatives in developing locations. One should ask how a project will affect long-standing community relationships or hierarchies; an example is the effect of new sources of wealth or prestige deriving from aquaculture (Ruddle, 1993), as happens with Polynesian pearl culture. Family relationships and the male-female household division of labor can be seriously disrupted, for example, when the project requires full-time employees or when only casual or seasonal workers are available (Smith and Peterson, 1982).

Community hierarchies and fundamental belief systems also can weigh heavily on the success of small- or large-scale development projects. Leaders (e.g., village chiefs), formal town councils, and other such groups should be identified and informed, and they must support the project if it is to be accepted by the community at large (Roy, 1985). Traditional knowledge vested in a community's values, customs, practices, and skills should be assessed to determine their compatibility with project goals and technology (Smith, 1984). Conflicts in any of these areas can undermine community support and sustainability.

Finally, many authors have described the difficulties of transferring a new technology to rural developing communities (Smith and Peterson, 1982; Roy, 1985; Ruddle, 1993). The general attributes of a technological innovation—i.e., complexity, compatibility, advantage, testability, and observation—should be assessed and incorporated into the project plan (Ruddle, 1993). Aquacultural innovation can be a "two-edged sword": it has the potential to add to the general community welfare, but it may sometimes cause unanticipated harm to community values, structures, and institutions.

PLANNING AND REGULATION

Institutional Framework: National Policy Development and Planning

Aquaculture has a critical place in the future of world food security and economic development. The most important role for national governments is to plan and administer expansion in a sustainable fashion. It is the exclusive role of government to set the national agenda for aquacultural expansion, including the necessary administrative, promotional, regulatory, and enforcement frameworks for its realization (Table 9.7). If the national government is hostile or indifferent to aquaculture, and more pointedly to a sustainable approach, the industry will not develop, or worse, the "boom and bust" cycles of shrimp culture, for example, will continue to occur.

Comprehensive public-sector development and planning for aquacultural policy has become more prevalent, though sustainability *per se* has not been a major component. The United States, Norway, Chile, China, Thailand, and Canada have dramatically expanded their industries through detailed development plans and

TABLE 9.7. Characteristics of a Successful and Sustainable Aquaculture Industry Development around the World

- Establishment of formal objectives that match industry imperatives and the socioeconomic setting
- Establishment of aquaculture as a national priority
- Formulation of an industry-driven research agenda
- Coordinated and continual interaction between public and private sectors on development activities
- Planning process with a feedback mechanism that allows the industry to incorporate the latest technology
- Establishment of new niches, markets, and roles for stimulating future industry expansion
- Mechanisms to identify and make available new production sites
- Procedures for resolving multiple-use conflicts
- Mechanisms to establish tenure or ownership of production sites
- Policy and regulations for ownership of aquaculture operations by foreign or multinational companies
- Coordination of local and national regulations, licenses, and permits
- Site-specific policy; legal, social and environmental infrastructure; market, technical support, and conditions needed to favor development

Source: Katz (1995a).

strong national aquacultural policies that make aquaculture a national priority (Katz, 1995a, 1995b). Where institutional efforts have not fostered growth, problems include insufficient linkage between aquaculture and other relevant policies, a preoccupation in the planning process with what is technically possible rather than with what is economically feasible and socially acceptable, and too limited and isolated a plan preparation process (UNDP, NMDC, and FAO, 1987). These general shortcomings were reiterated by a large regional study reviewing aquacultural sustainability and the environment (NACA Project Team, 1995). Governments wanting to foster sustainability should develop clear and achievable policies for aquacultural development, based on financial, social, and environmental sustainability (see earlier discussion of problems in the section "Comparing Visions for Aquacultural Development"). Moreover, these sustainability-based policies should be formed using a broad-based dialog among national agencies, nongovernmental organizations (NGOs), the private sector (particularly local farmers), and collaborating international institutions.

Generally, policy analysis provides alternative approaches that are used in decision making by planners and political officials. The process includes several critical components: problem definition; specification of evaluation criteria; generation of alternatives; specification of implementation procedures; and the identification of next steps, including procedures for monitoring, evaluation, and reassessment. The tools of modern governmental policy analysis vary from qualitative techniques, such as scenario building and surveys of people's attitudes and beliefs, to the very sophisticated quantitative methods prevalent in developed countries, such as cost-benefit analysis, linear programming, and modeling and simulation techniques (So et al., 1986; see also Chapter 7). In practice, the method of necessity and that often applied to aquaculture, is the most practical and simplistic approach using descriptive statistics and tabulation of costs.

Formulation of a national policy framework is followed by preparation of a plan, which includes detailed assessments of natural resources (e.g., land and water), technical resources, human resources, institutional/legal framework, and financial resources needed for development (Pillay, 1977, 1990). Staffing for a comprehensive sustainability planning effort should be multidisciplinary, and include persons knowledgeable in aquaculture, environmental science, economics and marketing, business and finance, public administration, and sociocultural issues. Moreover, the plan should utilize holistic, integrated, and existing information from other planning or development activities (e.g., fisheries, agriculture, environmental protection, and coastal-zone management), to minimize the need for new studies and keep cost and complexity down (Corbin and Gibson, 1979).

In general, a plan should describe the state or national goals for aquaculture and outline actions for achieving them, including time frame, costs, and implementing agencies. There are many approaches for modern public-sector planning; however, there are seven important considerations for any planning process (So et al., 1986):

1. *Organizational Location and Clientele.* Placement and subject matter need to be mission- and clientele-relevant.

2. *Purpose.* Changes and the actions to achieve goals need to be addressed.
3. *Openness.* There must be latitude for input and review by other agencies and the public.
4. *Time Horizon.* A definite time frame with specific actions is required.
5. *Scope.* The planning process must be broad.
6. *Specificity.* The degree of detail in requirements to be followed should be clear.
7. *Flexibility.* There must be an ability to meet changing conditions and degree of responsiveness.

Upon completion, a lead agency should be designated and given the responsibility and accountability for the aquaculture plan implementation. It is essential to place the implementation function as high in the lead agency as possible—at the division level, in the office of a cabinet-level official, or even with the president of a country—to give the lead agency the activity resources, stature, and political power needed to work aggressively with other departments and local governments (Corbin and Young, 1988; Tobin, 1992).

Regulation for Sustainability

A Mandatory Environmental Assessment Process. Uncontrolled and unregulated aquacultural development can pollute the environment, disrupt rural communities, and encourage overuse of natural resources. Mandatory assessments of preconstruction project impacts and contemporary approaches to regulation and enforcement are effective in managing critical aspects of aquacultural expansion in many developed countries and are increasingly being applied in the developing world (FAO and NACA, 1995). To achieve long-term, sustainable aquacultural farming, potential problems and concerns expressed by other government agencies and the public must be anticipated in the planning and design phases, and be eliminated or mitigated.

The United States pioneered mandatory environmental assessments in the early 1970s, through the National Environmental Policy Act and its subsequent amendments. Congress mandated all federal agencies to prepare detailed assessments on proposals for legislation and other major actions that significantly affected the quality of the human environment. State governments passed similar laws that encompassed most projects involving publicly managed resources (Rona, 1988).

The purpose of an Environmental Impact Assessment or Statement (EIA/EIS) is to provide a consistent, open process to evaluate the environmental consequences of development actions. An EIA/EIS in the United States generates advisory opinions and report findings, and offers recommendations to regulatory agencies and developers. Strictly speaking, it is not a decision-making process; rather, it is a tool in the balancing of environmental effects with other consequences, allowing a political decision (Carpenter and Maragos, 1989). Its users are government decision makers who issue various permits and approvals, and most importantly the general public.

The preparation and review of an EIA/EIS may vary, but several common aspects should be noted. Government may legally exempt certain development activities from the EIA/EIS requirement because they are deemed routine and environmentally insignificant. Preparation of an EIA/EIS in the United States is preceded by a limited assessment study to determine whether a full-scale EIA/EIS is necessary. If no significant adverse impacts are anticipated, the accepting EIA/EIS agency will issue a Finding of No Significant Impact, and the initial assessment will stand. When a comprehensive EIA/EIS is required, it is an expanded document describing areas of potential degradation identified during the initial review (Rona, 1988).

An important aspect widely adopted is "scoping," where all the potentially affected public- and private-sector agencies, groups, and individuals are invited to a meeting prior to an EIA/EIS preparation. Here they describe their issues and concerns, setting content guidelines for the assessment. Positive aspects of the project should be addressed, as well as the potential negative consequences (Pillay, 1990). The content of each EIA/EIS document should be tailored to fit a specific project situation. However, the following categories of information are usually found: (1) purpose of the project; (2) alternatives to the project and mitigation measures; (3) existing environmental conditions; (4) anticipated changes to the environment from the project; (5) impacts on human health and welfare; (6) secondary or indirect community impacts; (7) sustainability of the activity, including tradeoffs, options, and irreversible consequences; and (8) benefits of the proposed project (Carpenter and Maragos, 1989).

Disclosure to governmental agencies and the general public is made for purposes of receiving specific comments. The process brings out problems and allows the project to be modified in consideration of these concerns. To illustrate the concept, Hawaii's aquaculture projects have been asked by permitting agencies, based on the EIA/EIS process, to change stocking densities, reconfigure production systems, add fail-safe equipment and procedures, reduce water discharge, add noise-suppression devices, and redesign the physical layout of a farm.

Application of the EIA process and related legislation in other parts of the world has not necessarily followed the U.S. law precisely (Pillay, 1992). Different cultures, circumstances, and legal systems, for example, have shaped laws to fit the social, economic, constitutional, and technological frameworks of their country. Terms such as EIA and EIS are not comparable from nation to nation. In most countries, an EIA includes the technical information collection and analyses, while an EIS is the summary of results and recommendations to an agency (Pillay, 1992). In recent years, most multilateral and bilateral assistance agencies have begun to require environmental assessments for their large-scale development projects. The Asian Development Bank has perhaps the most comprehensive set of environmental assessment guidelines to encourage sustainable development and the most experience with implementing large-scale projects (Office of the Environment, 1991).

Government Permit Approaches. Governments the world over have many regulations for natural resource-based businesses like aquaculture. Commonly, a country's legal system is directed at managing basic activities such as land use, water use, environmental protection, pollution, public health, agriculture, or fisheries. In general, environmental and land-use regulations affecting aquaculture are intended to curb adverse impacts by (1) restricting available sites, (2) encouraging avoidance of environmentally sensitive areas, (3) specifying the type of facilities that may be constructed, and (4) specifying how facilities may be operated to reduce or mitigate impacts (Rubino and Wilson, 1993).

Water Quality. In the area of aquatic pollution, U.S. laws and regulations are often used as models for other countries. The National Clean Water Act of 1977 mandates a goal of achieving "fishable and swimable" waters for the nation. The law includes a mechanism, the National Pollutant Discharge Elimination System (NPDES), to manage via a permit system, fixed point-source discharges into natural waters. NPDES permits specify standards for effluents, thus limiting pollutant loads, and require monitoring reports to be submitted to government officials along with water-quality data to provide accountability (Rona, 1988). The actual program administration is delegated to the individual states, which may institute more stringent standards.

Under the program, point-source pollutants are classified as toxic, conventional (e.g., nutrients), or nonconventional (e.g., chemicals). Aquacultural effluents are classified as low-risk pollutants, and in fact the responsible federal agency, the U.S. Environmental Protection Agency, has yet to develop national standards for the industry (Bastian, 1991). Notably, the federal law includes exemptions for small aquacultural farms producing less than 9090 kg harvest weight of cold-water organisms in ponds, raceways, and other similar structures, or 45,454 kg of warm-water animals each year (Rubino and Wilson, 1993).

An additional regulatory mechanism was established by the Clean Water Act allowing states to designate receiving-water standards (i.e., limits of change for temperature, nutrients, or turbidity) as a result of a point-source discharge. Ideally, the requisite standards are determined from sound, scientific baseline environmental studies that describe existing conditions. Recognizing that administrative flexibility was needed, an adjunct permit mechanism, the Zone of Mixing Permit, was devised, particularly for exposed ocean coasts, to allow for violation of receiving-water standards over a specified area. This approach recognized that certain relatively benign discharges (e.g., from aquacultural farms) can and should be assimilated by the environment through dilution and mixing (Bay, 1995).

These regulatory concepts—(1) control of the discharge, (2) management of receiving-water-quality standards, and (3) selective granting of variances for specific areas—are the basis for a dramatic improvement in water quality in the United States. This permitting system can be utilized to properly site new aquacultural farms, but importantly, it is possible to close down farms that are not meeting the requirements, since permits must be renewed every 5 years. While these positive results are laudable, the negative side of regulation has been a significantly increased cost of doing business and, in the case of aquaculture, a severe restriction on industry expansion due to cost and availability of suitable sites.

Site-Specific Zoning. Major environmental and social problems have occurred when aquacultural farms were located in areas with incompatible activities. Conflicts between aquacultural farmers and other beneficiaries of resources could be reduced through ecologically and sociologically "considerate" zoning for specific aquacultural uses. Many recent authors have suggested government-initiated zoning for managing sustainable aquacultural expansion, but there are few examples of its use (Pillay, 1992; Aiken and Sinclair, 1995).

Zoning means dividing an area into districts, and regulating the use of the land or nearshore waters within these districts to promote the general welfare of the community. In developed countries, the process of zoning results in specific plans with associated use requirements and usually is a part of the master planning, designating what activities can occur and where (So et al., 1979). Zoning processes need public input and flexibility to administer in the "real world"; normally they include a variance procedure to allow a nonconforming use to occur under special situations and circumstances (Smith, 1983).

Perhaps the best example to date of identifying appropriate zones for encouraging sustainable aquaculture is from Norway's LENKA Project. This comprehensive, integrated site assessment and designation effort partitioned the entire coastal zone into specific areas designated suitable for the net-pen culture of salmonids. Parameters measured in the zoning process included environmental properties, existing uses, infrastructure, and special community concerns. In addition, a method of assessing the carrying capacity—the quantity of aquacultural activity possible in an area without causing damage to the environment—of LENKA zones was devised based on two main considerations: (1) the environmental impact from mariculture, and (2) the marine environment's impact on the cultured organism (Ibrekk and Elvestad, 1990).

Species Importation Issues. The deliberate and accidental releases of marine, brackish-water and freshwater organisms by aquaculture has led to significant ecological and economic consequences around the world. Indiscriminate, uncontrolled movement of organisms outside their natural range—i.e., exotic-species introductions—have resulted in serious ecological changes, negative impacts on native species, introduction of diseases and pest organisms, and detrimental economic impacts on other industries (Chew, 1990; Carlton, 1992; Lightner et al., 1992; Pillay, 1992; Rosenfield and Mann, 1992; Thorgaard and Allen, 1992). In contrast, introductions have also led to the establishment of aquaculture industries that resulted in positive economic impacts for their surrounding communities (OTA, 1993b).

The tradeoffs and risks involved in utilizing nonnative species are varied. Aquaculturists are usually seeking species with improved production characteristics or market value. Adapting a carefully chosen exotic species to a new site may be easier and less costly than solving the biotechnical problems and developing commercial methods for untried local species. Governments, as stewards of natural resources and economic prosperity, are increasingly regulating species importations to reduce the risk of harmful introductions. See also Chapter 4.

Two widely adapted approaches are (1) the International Council for the Exploration of the Seas (ICES) Code of Practice, a voluntary protocol for aquatic species introductions that relies on a quarantine procedure and uses first-generation progeny (Sindermann, 1986, 1992); and (2) the American Fishery Society (AFS) Protocol, where an exotic species has to pass a five-point evaluation that begins with an extensive review of the published literature (Kohler and Courtenay, 1986). The preliminary review and quarantine processes for a permit to import an exotic species have been incorporated into some state laws in the United States. The ICES quarantine protocols have been adopted by many nations and represent a rigorous process for risk assessment and management.

Many states in the United States use a system of "clean" or "dirty" lists to administer aquatic species introductions, the primary concern being escape and environmental impacts (OTA, 1993b). Hawaii has one of the more rigorous processes and utilizes lists of prohibited and approved species, such that any species not listed is automatically prohibited. The application procedure to list a species includes preparing a detailed application, gathering advisory opinions from experts, allowing review by natural resource agencies, acquiring input from the public at large, and getting approval from the Board of Agriculture (Davidson et al., 1992; Olin, 1993). Special attention is paid to the origin of healthy stock; i.e., the state veterinarian is required to certify the health of the stock from the point of origin or provide local inspection upon arrival, and containment engineering (particularly controlled disposal of wastewater) is required at the receiving facility.

Enforcement Approaches. Regulation as a management tool is only as effective as the ability of government to monitor and enforce compliance. While developing countries may have a limited framework of environmental laws and regulations, developed countries have ongoing problems with the cost and effectiveness of mandated enforcement activities.

Command and Control. With the command-and-control approach, a set of codified laws and regulations dictates what can and cannot be done with and in the environment, giving specified penalties for noncompliance. Compliance is generally monitored by an agency through required reports, inspection visits, or response to complaints from the general public. Countries with large, relatively undeveloped natural resources have difficulties in directly monitoring and enforcing abuses, including those from aquacultural operations. Environmental damage may occur through intentional, negligent or accidental conduct. If legal action is required, options include: civil actions (e.g., in cases of property damage); criminal actions (e.g., in purposeful harm to human life); or application of administrative procedures such as refusal to grant or renew a permit (e.g., failure to comply with monitoring procedures; Van Houtte, 1995).

Both developed and developing countries have used a farm licensing system for oversight of aquacultural development (FAO and NACA, 1995). The ability to grant licenses can provide governments with much influence over aquacultural development, including numbers of farms, farm location, specific operating conditions, and record keeping and reporting. Legally, it is an offense to operate an aquaculture enterprise without a license or in violation of the conditions on a license. Punishment for breaking the law could include a substantial financial penalty, suspension or termination of the license, and/or actual shutdown (Howarth, 1995).

Self-Regulation. When enforcement must encompass large expanses of resources and numerous resource users, and where it is difficult to identify abusers, other more

TABLE 9.8. Washington Fish Growers Association Best Management Practices Categories of Information for Net Pens

Site Selection

- Compliance with all federal, state, and local laws
- Preapplication agency consultation
- Good-neighbor policy of communication and consultation
- Aesthetic considerations

Operation

- Introduction of fish stocks
- Feeds and feeding
- Mortality removal and disposal
- Net cleaning and facility maintenance
- Predator control
- Harvesting
- Medication and chemical application, spill prevention, and chemical storage
- Waste collection and disposal
- Environmental monitoring and reporting
- Navigation aids, emergency plans, and safety
- Aesthetic and noise considerations
- Employee education
- Public education and involvement

Source: Rubino and Wilson (1993).

effective and lower-cost regulatory measures must be sought. Industry self-regulation or voluntary compliance to the law is gaining in popularity because it reduces the necessity of governmental oversight. Underpinning this concept is the awareness of shared or community responsibility for the health of the environment and the necessity of a strong joint industry-government community education effort.

Two examples of self-regulation for aquaculture are the Best Management Practices (BMPs) to reduce waste and increase operational efficiencies, and the industry-generated and enforced Codes of Practice (COPs). Both approaches encourage improved, uniform management procedures across the industry. BMPs can be developed by federal or state agencies or by the industry itself, as voluntary standards for such aspects as site selection and farm operation (Table 9.8). COPs are similar to BMPs but usually are developed by industry associations (e.g., the Irish Salmon Growers Association's "Good Farmers, Good Neighbors" COP) to cover their entire salmon production and distribution process. COPs have the added advantage of using their existence to produce quality products by environmentally sound practices as a marketing advantage. In general, BMPs and COPs are "policed" by the industry with violators losing the privileges of belonging to the industry association.

However, government has also been known to incorporate BMPs as part of permit or licensing conditions (Bay, 1995).

Market-Based Approaches. Policy makers are realizing that if government's environmental goals can be measured, market-oriented approaches can efficiently replace many regulations (Roodman, 1996). Market-based incentives operate through direct and indirect financial rewards or penalties. General examples of market-based approaches to compliance that may be relevant to aquaculture include tax reductions, tradable pollution permits and credits, and user fees. Use of the tax code to encourage environmental conservation is widely applied around the world and includes tax incentives, such as property tax reductions for specified operating procedures or credits for expenses incurred in improving environmental management (Roodman, 1996). Tradable use permits or credits to pollute the environment create a limited market for access to a geographic area, where a government has set limits on the amount of use or pollution allowable. An example of this would be issuing permits to allow the use of a large coastal bay for discharge of effluents from a finite number of permitted shrimp farms (Van Houtte, 1995). User charges, on the other hand, pass the public costs of keeping the environment clean and protected on to the industries that pollute, usually in proportion to the amount of degradation they cause.

In practice, some types of government fees are incentives or disincentives to change polluting behaviors, while others are income generators to support monitoring or the costs of administration (Van Houtte, 1995). Caution should be used in applying these concepts of regulation and enforcement to developing countries in a form unmodified for each country's particular sociocultural conditions. One concern is that market mechanisms could take proportionally more money from poor people than from the middle and upper classes (Roodman, 1996). Moreover, approaches to regulation and enforcement in developing countries should be compatible with institutional arrangements, cultural practices, and local customs because they will gain greater public acceptance (Johannes, 1978).

PROACTIVE DEVELOPMENT APPROACHES FOR SUSTAINABILITY

Government in its proactive role of stimulating economic development can provide strong leadership and direction for achieving sustainable aquacultural development. Virtually all countries that are actively encouraging aquaculture provide various forms of direct assistance and financial incentives to foster expansion (Katz, 1995a, 1995b). Administration of these policies and programs with an underlying national goal of having a sustainable industry would provide a tremendous opportunity for eliminating or mitigating problems before they occur and for supporting viable species and technologies (Table 9.9).

TABLE 9.9. Development Tools for Sustainable Aquacultural Development

Financial incentives:
Grants
Loans
Interest subsidies
Equity financing
Tax incentives
Nonfinancial assistance:
Site selection
Facilitation of licenses, regulations, permits
Market/marketing studies
Industry-driven research-and-development services
Training and workshops
Production and engineering extension services
Veterinarian and diagnostic services

Source: National Association of State Development Agencies (NASDA) (1993).

Linking Government Assistance to Sustainability

Applied Research. Research, particularly applied research, has been a driving force in the rapid expansion of global aquaculture over the past 20 years. Ongoing scientific exploration is necessary for development of new products, industry roles, and production technologies as the field unfolds and markets change. Aquaculture has prospered in countries where research has been closely linked to industry needs and participation in setting research agendas (e.g., Japan, Chile, Norway, and Taiwan; Katz, 1995a, 1995b).

Industry-driven, publicly funded research may or may not address issues relevant to sustainability. For example, fish feed and nutrition research could focus on maximum growth, with minimal protein and maximum feed conversion ratios, but not be concerned with pollution potential and environmental sustainability (Pillay, 1992; Jory, 1995). However, due to recent concerns over pollution from feed-laden effluents of salmon farms, researchers from Denmark have added objectives to national programs to improve feed digestibility and reduce the release of pollution-causing nutrients into the environment (Jensen and Alsted, 1990).

It is important for both government policy makers and the industry to realize that national and international research must be focused on questions affecting sustainability, if the knowledge to plan and manage farms successfully is going to be developed. An effective research agenda for sustainability will require the joint setting of objectives by government and industry; an example of this is Japan's industry-driven, applied-research programs carried out jointly at numerous centers (Katz, 1995b). This cooperative approach is used by the U.S. Department of Agriculture's five Regional Aquaculture Centers, which administer federal funding for cooperative research and demonstration and extension projects targeted at commercial opportunities in each region. Opportunities abound to focus applied research in both developing and developed countries on improving the resource use efficiency of production systems, as well as on forecasting and mitigating the environmental impacts of siting farms.

A recent regional study of environmental assessment and management of aquaculture identified many research areas that need funding to support further policy development, including the assimilative capacities of target environments, processes leading to eutrophication, allocation of natural resources, effluent standards and environmental monitoring systems, use of integrated agriculture-aquaculture systems and polyculture systems, rehabilitation of abandoned culture areas, environmental effects of aquaculture drugs and chemicals, and water-quality management (FAO and NACA, 1995).

Development Assistance. Government provides targeted, hands-on assistance programs to the aquaculture industry to encourage private investment, reduce the risk of failure, and expand the economy. These essential economic development efforts can also be implemented to encourage sustainable farming and to implement a country's sustainable aquaculture policy. Two development-assistance approaches have received widespread use, mainly in the United States but also elsewhere around the world, though not necessarily to promote sustainability: (1) provision of selected business services and technical assistance, and (2) the use of quasi-public agencies to render comprehensive assistance for new business development in a particular geographic area (NASDA, 1991).

Start-up aquaculture projects, whether for-profit businesses or family- or community-based subsistence farming, generally rely on government for substantial technical assistance. These advisory efforts may include site selection and construction; obtaining stock; feed management and composition; water-quality and disease management; and processing, handling, and marketing of the final product. Generally, these skilled extension and advisory personnel are housed at universities or governmental agencies, but they are not as universally available as would be desirable.

Extension services play a critical role in government's aquacultural research efforts as the principal delivery system for results. Agents travel to the countryside and provide hands-on training and transfer technology to farmers; they also identify research problems for the national scientific community (Davis, 1992). Educating the next generation of extension workers, in light of rapidly evolving methods and technologies for sustainable production, is a critical job for universities around the world in the next century (FAO and NACA, 1995). This job is made even more difficult by the diversity of aquacultural species and technologies, and the profound differences in social and economic situations within rural communities, as well as the competition for limited resources by national aquacultural extension service and research programs.

Direct governmental assistance to the industry can also take an emergency, problem-solving role. For example, Thailand's rapid growth of marine shrimp farming is causing serious environmental problems along the coast. The Thai Department of Fisheries is undertaking a large construction project to build seawater irrigation systems and excavate river mouths to improve water exchange. In addition, the government is establishing a laboratory in every shrimp-growing area to detect antibiotic residues in the pond-grown product (Kongkeo, 1995).

Many countries approach aquacultural development on a regional basis, when issues (e.g., watershed management) extend beyond the political boundaries and jurisdiction of a single entity's capability to manage them. Economic development is often approached on a regional basis through quasi-governmental regional organizations that assist new industries, like aquaculture. Orienting these agencies toward encouraging sustainable species and technologies is a policy decision made by the government that creates them.

Fundacion Chile (FC) is an example of a quasi-governmental organization that has successfully spearheaded rapid development of aquaculture in remote coastal areas (Katz, 1995a, 1995b). The FC was developed as a nonprofit organization in 1976 through initial funding from the Chilean government and the private sector, to transfer new, economically beneficial technologies, including those of aquaculture. To be self-sufficient, the FC was allowed to charge fees for services or earn revenues from the commercial enterprises it established. Aquaculture of Pacific oysters and salmon initially became part of the FC portfolio through establishment of two subsidiaries. The FC also facilitated private investment by preparing plans, conducting feasibility studies, solving technical problems, building facilities, training personnel, and privatizing FC-developed operations.

Another illustration is Scotland's Highlands and Island Development Board (HIDB), which was created by the British Parliament in 1965 to revitalize the economically depressed Orkney and Shetland Islands region. Salmon pen farming was identified as a tool for business development and jobs. A long-term program of public-funded research and pilot demonstration projects was followed by grants, loans, technical assistance, and training through various agencies and universities (Katz, 1995a, 1995b). After the industry was established, public assistance to salmon farmers continued at reduced levels and with more emphasis on applied research, training and market development.

The transfer of new technology through a cooperative effort between an international assistance agency and a private contractor as the transfer agent is another approach. Undertaking technology transfer substantially depends on the need or desire for the aquacultured products, on the suitability of natural resources and complementary farming factors, and on basic resident aquacultural knowledge. The speed of transfer largely depends on the language skills and cultural sensitivity of the transfer personnel, which are gained through repeated attempts at communications, training, and personal relationship building. The success of Hawaii's private, non-profit Oceanic Institute in technology transfer for milkfish and mullet technologies to Indonesia and Egypt, respectively, exemplify the importance of these observations (Lee and Lee, 1995).

Investment Incentives. Governments frequently utilize financial incentives to entice private investment in selected industries (NASDA, 1991). With aquaculture, both developed and developing countries have applied a mixture of incentives, including grants, low-interest loans, free land, permit exemptions, and tax breaks to develop their aquacultural potentials (UNDP and NMDC and FAO, 1987; Katz, 1995a, 1995b). In developing countries where aquaculture is a priority, principal sources of financing comes through loans from multilateral assistance agencies (e.g., the World Bank, the Asian Development Bank, and investments by multinational corporations; UNDP, NMDC, and FAO, 1987; Nash, 1987). Clearly, if government tied the availability of these incentives to its long-range goals and policies for sustainability, it would induce development interests to comply.

Grants are perhaps the most desirable direct financial incentive because there is no payback obligations by farmers. Low-interest public loans, in contrast, require repayment, though the terms and conditions may be generous. Small start-up firms without an operating track record seek government grants or loans because of their inability to obtain credit from commercial lenders. In addition, start-up sustainable aquaculture projects should have available to them government grants owing to the greater risks and limited private investment sources for such projects (Angus, 1990). It should be emphasized that there are many intrinsic and extrinsic risks to aquacultural farming, the most important of these being in fostering sustainability is good management (Sandy, 1990).

All businesses are required to pay some form of taxes to the government. Abatements, exemptions, and moratoriums on taxes are other types of financial inducements offered to aquacultural farmers. Specifically, opportunities to assist a new business using tax codes are quite varied, including offsets for capital investments or hiring local workers, sales-tax "holidays" to allow the farm to generate a positive cash flow, offsets for new equipment, and exemptions to encourage exporting.

Other tangible incentives, such as free land or regulatory relief (e.g., permit exemptions) can also work well to foster sustainability. Land could be made available in remote sites that need economic development but also have the carrying capacity for aquatic farming. Likewise, permit exemptions could be allowed if certain BMPs were employed by the farm. With the government's abilities to put conditions on any agreement it has with an industry, it is in a strong position to develop aquatic farming that is sustainable over the long term.

Aquaculture Parks

Government has two major tools to manage and control population and development: land-use zoning (previously discussed) and strategic use of capital improvement projects (CIPs) through the building of public infrastructure. The principle behind the use of CIPs is that population and business growth will tend to follow the construction of public roads, utilities, and so on as communities develop (So et al., 1986).

Developed-country governments have strategically used CIP expenditures to construct industrial parks, where businesses (especially high-technology ones) can locate to both their benefit and the economic benefit of the nearby community (NASDA, 1991). The concept is based on public investment in the park for roads, buildings, utilities, and so on, with the payback coming through the business investments and jobs generated. An additional benefit of building special-purpose parks is the grouping of compatible business activities. Thus, a well-planned aquacultural park prevents potential environmental and other conflicts caused by scattering activities or locating incompatible activities in, for example, pristine natural areas.

Land-based aquacultural parks have been constructed in Japan (for marine species), Singapore (for freshwater tropical fish), and the United States (for marine species). In the state of Hawaii developed the 830-acre Natural Energy Laboratory of Hawaii Authority (NELHA) with public funds to allow expansion of ocean-related research and commercial businesses, including mariculture (NELHA, 1996). Public funds leveled the site and put in roads, utilities, and large-capacity pipes, which bring up 8°C, pathogen-free, deep ocean water for commercial mariculture and other purposes (ADP, 1993). Companies at NELHA can start up operations quickly and with lower initial costs because of master permitting; that is activities have been approved, and the infrastructure is in place. They pay back this public investment through direct revenues (i.e., land lease rents and fees), indirect revenues (i.e., private infrastructure investment), purchase of equipment and supplies, salaries paid, and taxes. These expenditures have a large multiplier effect as they circulate in the commerce of the surrounding community.

Community-Based Development

Modern society has a highly developed technology base and division of labor, a high value placed on education and literacy, economically rational planning, and efficient external channels of communication. In contrast, a traditional society possesses relatively less developed technology, lower levels of education and literacy, planning based predominantly on social rather than economic considerations, and limited external communication (Blakely and Hrusa, 1989). Traditional communities may be found in rural areas of both developed and developing countries, and aquaculture has been used by governments as a tool to assist their economic, social, and nutritional needs.

Rather than encouraging large-scale, unsustainable, high-technology, commercial systems in a "one size fits all" mentality, many governments have adopted a small-scale community-based approach, which builds on the capacity of local people and their cultural values, utilizes natural and community resources in ways that are sustainable, and reinvests in the community for widespread benefit (Skladany and Bailey, 1994). Equally important in this approach, the social and cultural values of a community are reinforced and maintained (Burgess and Hrusa, 1995).

Governments can directly initiate community-based projects or act as catalysts to help mobilize community groups. Alternatively, communities may informally organize into cooperatives or, as in the United States, formally establish nonprofit, nongovernmental associations to mobilize resources (Burgess and Hrusa, 1995). This grassroots type of planning and development is active in various forms in Africa, Central and South America, and Asia and often leads to integrated agriculture-aquaculture production systems that can be highly resource-efficient (Brummett and Noble, 1995). These highly sustainable, small-scale subsistence and for-profit enterprises are supported through training programs, technical assistance to solve problems (e.g., disease), and financing programs (Blakely and Hrusa, 1989; Burgess and Hrusa, 1995).

CONCLUSIONS AND PROGNOSIS

Conclusions

Successful sustainable aquaculture is perpetually meeting the demands for food production at socially acceptable economic and environmental costs. Government has a central role in planning and managing sustainable development for society, from high-technology, high-yield, export production to small-scale, integrated, subsistence production. Most governments will have mixed aquacultural goals and strongly desire the benefits (i.e., foreign exchange and investments) of a large-scale export orientation, while having a segment of the population that would be greatly assisted by small-scale production for local consumption. Clearly, both goals should be accommodated in any sustainable aquaculture vision, and indeed, there is economic and social synergy stability in having a wide spectrum of aquacultural purposes, scales, and species under cultivation.

More widely available remote-sensing technologies combined with low-cost GIS applications will make future large-scale natural resource planning and management more effective in both developing and developed countries. Capacity building of expertise must be a priority, as the rapid evolution of technology and applications is outstripping the pool of technicians and managers needed for skilled operation.

The information needs and tools for local or project-specific site assessment, particularly environmental concerns, are becoming well-known (Carpenter and Maragos, 1989). Establishing baseline information will continue to be an important consideration in evaluating environmental changes. Likewise, more persons trained in marine ecology and environmental science for aquaculture project assessment will be needed in developing countries to take full advantage of the extensive natural resources available. More attention from the academic and private consulting communities is needed in evaluating the economic and sociocultural characteristics of a particular site and in forecasting the impacts of development.

Sustainability requires a holistic approach to considering resource needs and potentials, but aquaculture should also be planned and integrated with the larger natural-resource management, environmental-quality, economic development, and sociocultural goals of a country. Again, it will be challenging to find the expertise that can manage the integrated approach required to make sustainable aquaculture successful.

Regulation offers one means for directly guiding aquaculture into a sustainable mode of development. Arguably, developed countries (e.g., Norway and the United States) that have incorporated both environmental and aquacultural consciousness into their laws are already well along this path. Developing countries, (e.g., those in Southeast Asia) that are beginning to adopt mandatory environmental assessments, water-quality standards, species-importation procedures, and land- and water-use zoning to manage development should adapt these concepts to their political, economic, social, and cultural settings.

It is important to emphasize that the use of mandatory preconstruction environmental assessments will be an essential tool everywhere, if aquaculture is going to be sustainable. Few developing countries use this approach at the present time, and model legislation for ready adaptation in particular countries would be helpful (FAO and NACA, 1995). National governments and the development banks (e.g., Norway and the Asian Development Bank) will have to be in the forefront in conducting and publicizing these types of studies.

Workable regulatory approaches for aquacultural sustainability require flexibility in their application to the real world, and use of techniques (e.g., seasonal standards for places with annual monsoon rains) should be explored. As a new technology, aquaculture faces problems being subjected to a long-standing legal framework that developed without considering its needs, but these may be addressed by aquaculture-specific laws and regulations (Howarth, 1995). Furthermore, the value of using postconstruction assessments and standards for review of existing projects—either have the projects reengineered to adopt sustainable management techniques or to have them shut down—should not be overlooked.

The cutting edge of new enforcement approaches includes the use of voluntary compliance through self-regulation (e.g., Codes of Practice) and market forces (e.g., tax incentives). Some countries will require use of straightforward laws and regulations with heavy penalties for violators, because the industry does not have the necessary maturity, uniformity of production technology, and cohesiveness to apply the more cost-effective, less-governmental approaches.

Proactive approaches to expanding aquatic farming can also lead to focusing government and industry on sustainability. Applied research on technologies to support sustainability is a critical need, since conceptually these may differ from the common commercial aquacultural approach of using resources to obtain the highest yield for the least cost. Industry participation, in setting the research agenda, is critical to funding commercially meaningful studies. Likewise, extension personnel must be proponents of sustainable management practices and train successful farmers, who can then demonstrate the techniques to the wider community in keeping with the principles of technology transfer (Ruddle, 1993; Lee and Lee, 1995).

Proactive approaches also include financial and other incentives to encourage private investment in sustainable farms and farm management practices. Government has broad power and influence to foster beneficial economic, social, and cultural outcomes. Encouraging parks and community-based projects that embrace sustainability will require significant public investment in targeted capital improvements, technical assistance, and start-up loan funds, but tying these incentives to sustainable aquacultural farming can pay substantial, long-term rewards.

Prognosis

Once there is recognition that certain approaches can achieve sustainable aquacultural development, their successful use will depend largely on the following factors:

1. Government leaders must have the political will to apply the concepts for the long-term good of the environment, the economy, and the society.
2. Financial and technical resources must be available to allow for purchasing, organization, and the administration of more sophisticated planning and management approaches;
3. There must be extension personnel and public officials who understand the national goal of sustainability and can educate communities and companies in how to cost-effectively achieve the desired results.

In practice, sustainable aquacultural development does not strive to maximize any single result—whether it be environmental quality, economic return, jobs, or number of farms. Instead it strives to achieve a long-term balance among outcomes. Profit is but one of a host of business, environmental, and community objectives that will contribute to the long-term sustainability of any given project. Seeking to balance the objectives for aquacultural development leads to two central questions: What is the acceptable balance for the environmental, economic, and sociocultural costs of aquatic farming, and who ultimately judges the acceptability?

Achieving sustainability in aquacultural farming is about managing change brought about by a multitude of internalities and externalities (Bardach, 1995). The balance needed for any project is in reality a "moving target" that changes over time and from one generation to another. Continuous evaluation, adaptation, and managerial responsiveness are necessary factors in achieving success. More specifically, judging the balance to be achieved is the purview of society, and the target will change according to societal needs and consensus.

Sustainability targets should not be developed unilaterally by government, but government must involve the public—the stakeholders—in the process of setting goals and evaluating results. This participatory approach is being widely used in developed countries, and sophisticated techniques for facilitating public input into governmental decision making are finding application in developing countries (FAO and NACA, 1995). Improvement and wider application of these community- and place-based collaborative planning and management techniques will be essential to achieving sustainable aquacultural development now and into the next century.

In the final analysis, society judges whether the costs and benefits resulting from aquaculture are acceptable and whether sustainability has been achieved. Developing countries, in particular, must determine whether they now have the "luxury" of incorporating the concept of aquacultural sustainability into meeting their urgent needs for food, investment, and foreign exchange.

REFERENCES

Ackefors, H., and K. Grip, 1995. The Swedish model for coastal zone management, Report 4455. Swedish Environmental Protection Agency, Stockholm. 83 p.

ADP (Aquaculture Development Program). 1993. Hawaii's future in aquaculture strategy for the blue revolution. Aquaculture Development Program, Department of Land and Natural Resources, for the state of Hawaii, Honolulu. 124 p.

AFS. 1990. American Fisheries Society position paper: commercial aquaculture. American Fisheries Society, Bethesda, Md. 31 p.

Aiken, D., and M. Sinclair. 1995. From capture to culture, exploring the limits of marine productivity. World Aquacult., 26(3): 21–34.

Angus, I. 1990. Insurance and risk management. pp. 356–361, In: Proceedings of Aquaculture International. September 4–7, Vancouver, B.C. Heighway Ltd., Vancouver, B.C.

Austin, B. 1993. Environmental issues in the control of bacterial diseases of farmed fish. pp. 237–251, In: R. S. V. Pullin, H. Rosenthal, and J. L. Maclean (Eds.). Environment and aquaculture in developing countries. ICLARM/DGTZ, Manila, Philippines.

Baker, T. <tjb@unh.edu.> 1996. Landsat Pathfinder. National Aeronautics and Space Administration Humid Tropical Forest Inventory Project, table of contents. <http://pathfinder-www.sr.unh.edu/pathfinder1/index.html>.

Bardach, J. E. 1995. Aquaculture and sustainability. World Aquacult., 26: 1.

Bastian, R. K. 1991. Overview of federal regulations pertaining to aquaculture waste management and effluents. pp. 220–226, In: Proceedings of the National Workshop on Livestock, Poultry and Aquaculture Waste Management. July 29–31, Kansas City, Mo., American Society of Agricultural Engineers, St. Jo., Mi.

Bay, J. 1995. Permits and environmental requirements for aquaculture in Hawaii. Aquaculture Development Program, Department of Land and Natural Resources for the state of Hawaii, Honolulu. 76 p.

Beaujardiere, J. de La <delabeau@iniki.gsfc.nasa.gov.> 1996. Public use of remote sensing data. <http://camille.gsfc.nasa.gov/rsd/>. Revised May 8.

Blakely, D. R., and C. T. Hrusa. 1989. Inland aquaculture development handbook. Fishing News Books, Oxford, England. 184 p.

Brummett, R. E., and R. Noble. 1995. Aquaculture for African smallholders. *ICLARM Technical Report 46.* ICLARM, Manila, Philippines. 69 p.

Burgess, P., and C. T. Hrusa. 1995. Community-based economic development through backyard aquaculture. pp. 68–76, In: Sustainable Aquaculture '95. Proceedings of PACON 95 meeting. Pacific Congress of Marine Science and Technology, 11–14 June Ilikai Hotel, Honolulu. PACON, Honolulu.

Carlton, J. T. 1992. Dispersal of living organisms into aquatic ecosystems as mediated by aquaculture and fisheries activities. pp. 13–48, In: A. Rosenfield and R. Mann, (Eds.). Dispersal of living organisms into aquatic ecosystems. Maryland Sea Grant College Program, College Park, Md.

Carpenter, R. A., and J. E. Maragos. 1989. How to assess environmental impacts on tropical islands and coastal areas. East-West Center, Honolulu. 345 p.

Chew, K. K. 1990. Global bivalve shellfish introductions. World Aquacult., 21: 9–22.

Chua, T. E., and L. F. Scura. 1992. Integrated framework and methods for coastal area management. Proceedings of the Regional Workshop on Coastal Zone Planning and Management in Asean: Lessons Learned. April 28–30, 1992, ASEAN Conference Proceedings 12. International Center for Living Aquatic Resources Management, Manila, 169 p.

Conant, F., P. Rogers, M. Baumgardner, C. McKell, R. Sashmann, and P. Reining. 1983. Resource inventory and baseline study methods for developing countries. American Association for the Advancement of Science, New York. 539 p.

Corbin, J. S., and R. T. Gibson. 1979. Planning aquaculture development: the first time is always the hardest. Proc. World Maricult. Soc., 10: 22–27.

Corbin, J. S., and L. G. L. Young, 1988. Hawaii aquaculture resource planning and development—past and future. pp. 627–634, In: Proceedings of the Aquaculture International Congress and Exposition. Sept. 6–9, Vancouver, B.C. British Columbia Pavilion Corporation (BCP), Vancouver, B.C.

Corbin, J. S., and L. G. L. Young. 1995. Growing the aquarium products industry for Hawaii. Aquaculture Development Program, Department of Land and Natural Resources for the state of Hawaii, Honolulu. 35 p.

Davidson, J. R., J. A. Brock, and L. G. L. Young. 1992. Introduction of exotic species for aquaculture purposes. pp. 83–102, In: A. Rosenfield and R. Mann (Eds.). Dispersal of living organisms into aquatic ecosystems. Maryland Sea Grant College Program, College Park, Md.

Davis, J. T. 1992. Cooperative extension programmes in aquaculture (successes as a blueprint for the future). pp. 379–387, In: J. K. Wang and P. V. Dehadrai (Eds.). Aquaculture research needs for 2000 A.D. Oxford and IBH Publishing Co., New Delhi, India.

DeBoer, J. 1992. Building sustainable agricultural systems: economic and policy dimensions. pp. 51–74, In: Proceedings of the Regional Workshop on Sustainable Agricultural Development in Asia and the Pacific Region. June 15–19, Asian Development Bank and Winrock International, Manila, Philippines.

EDC DAAC<edcdaac@edcwww.cr.usgs.gov.> 1996. Welcome to the EDC DAAC. <http://edcwww.cr.usgs.gov/landdaac/landdaac.html.> Revised June 19, 1996.

FAO and NACA. 1995. Report on a regional study and workshop on the environmental assessment and management of aquaculture development, February 21–26, 1994, Bangkok, Thailand. FAO and NACA, Bangkok. 492 p.

Gowen, R. J., and H. Rosenthal. 1993. The environmental consequences of intensive coastal aquaculture in developed countries. pp. 102–115, In: R. S. V. Pullin, H. Rosenthal, and J. L. Maclean (Eds.). Environment and aquaculture in developing countries. ICLARM/DGTZ, Manila, Philippines.

Hall, A. <am.hallsw.edu.au@un.> 1996. Remote sensing and GIS. http://www.geog.unsw.edu.au/w̃eb/crgis/pics.html>.

Howarth, W. 1995. The essentials of aquaculture regulation. pp. 459–465, In: Report on a Regional Study and Workshop on the Environmental Assessment and Management of Aquaculture Development, February 21–26, Bangkok, Thailand, TCP/RAS/2253. FAO/NACA, Bangkok, Thailand.

Ibrekk, H. O., and S. Elvestad. 1990. The Norwegian experience. pp. 267–277, In: Proceedings of the Aquaculture International Congress, September 4–7, Vancouver, B.C. British Columbia Pavilion Corporation (BCP), Vancouver, B.C.

Ikeda, M., and F. Dobson (Eds.). 1995. Oceanographic applications of remote sensing. Lewis Publishers, Boca Raton, Fla. 512 p.

Jensen, P. M., and N. Alsted. 1990. How fish feed can be friendly to the environment. pp. 296–300, In: Proceedings of Aquaculture International, September 4–7, Vancouver, B.C. Heighway Ltd., Vancouver, B.C.

Johannes, R. E. 1978. Traditional marine conservation methods in Oceania and their demise. Annu. Rev. Ecol. Syst., 9: 349–363.

Jory, D. E. 1995. Marine shrimp farming development and current status, perspectives and the challenge of sustainability. pp. 35–44, In: Aquacult. Mag. Buyer's Guide '96. Aquaculture Magazine, Asheville, N.C.

Kam, S. P., J. N. Paw, and M. Loo. 1992. The use of remote sensing and geographic information systems in coastal management. pp. 107–132, In: Proceedings of the Regional Workshop on Coastal Zone Planning and Management in ASEAN: Lessons Learned, April 28–30. ASEAN Conference Proceedings 12.

Katz, A. 1995a. Aquaculture—international examples of success and failure and lessons for the United States. Associates in Rural Development, Burlington, Vt. (unpublished manuscript).

Katz, A. 1995b. Comparative strategies of aquaculture developed in select countries. p. 215, In: Sustainable Aquaculture '95, PACON International, June 11–14, Ilikai Hotel, Honolulu, Hawaii. PACON, Honolulu.

Kew, N. <nick.kew@pobox.com>. 1996. The Satellite Imagery FAQ.Contents. <http://www.geog.nott.ac.uk/remote/satfaq.html>. Revised June 16, 1996.

Kohler, C. C., and W. R. Courtenay, 1986. American Fisheries Society position on introductions of aquatic species. Fisheries, 11(2): 2–3.

Kongkeo, H. 1995. How Thailand made it to the top. Infofish Int., 1: 25–31.

Landesman, L. 1994. Negative impacts of coastal aquaculture development. World Aquacult., 25: 12–17.

Lee, C. S., and K. S. Lee. 1995. Implications of sustainable aquaculture in future technology transfer: the Oceanic Institute experience. pp. 234–241, In: Sustainable Aquaculture '95. Proceedings of PACON 95 meeting. Pacific Congress Marine Science and Technology, June 11–14, Ilikai Hotel, Honolulu, Hawaii. PACON, Honolulu.

Lightner, D. V., R. M. Redman, T. A. Bell, and R. B. Thurman. 1992. Geographic dispersion of the viruses IHHN, MBV and HPV as a consequence of transfers and introductions of penaeid shrimp to new regions for aquaculture purposes. pp. 155–174, In: A. Rosenfield and R. Mann, (Ed.). Dispersal of living organisms into aquatic ecosystems. Maryland Sea Grant College Program, College Park, Md.

Lyon, J. G., and J. McCarthy. 1995. Wetland and environmental applications of GIS. Lewis Publishers, Boca Raton, Fla.

McGuire, D. J., M. F. Goodchild, and D. W. Rhind. 1991. Geographical information systems. Longman Scientific and Technical, Essex, England.

NACA Project Team. 1995. Regional study and workshop on aquaculture sustainability and the environment, shrimp and carp aquaculture sustainability. Regional overview. RETA5534. ADB/NACA, Manila, Philippines. 98 p.

NASDA (National Association of State Development Agencies). 1983. Directory of incentives for business investments and development in the United States: a state-by-state guide. Urban Institute Press, Washington, D.C. 652 p.

NASDA (National Association of State Development Agencies). 1991. Directory of incentives for business investments and development in the United States: a state-by-state guide. Urban Institute Press, Washington, D.C. 778 p.

Nash, C. E. 1987. Aquaculture attracts increasing share of development aid. Fish Farm. Int., 6: 22–24.

NELHA (Natural Energy Laboratory of Hawaii Authority) <nelha@ILHawaii.net>. 1996. Welcome to the Natural Energy Laboratory of Hawaii, <http://bigisland.com/ñelha/>.

Newkirk, G. 1993. Do aquaculture projects fail by design? World Aquacult., 24(3): 13–18.

Office of the Environment. 1991. Environmental guidelines for selected agricultural and natural resources development projects. Asian Development Bank, Manila, Philippines. 115 p.

Olin, P. 1993. Importing live organisms to Hawaii: procedures and permitting. University of Hawaii Sea Grant College Program, Honolulu.

OTA (Office of Technology Assessment), U.S. Congress. 1993a. The future of remote sensing from space: civilian satellite systems and applications. OTA-ISC-558. U.S. Government Printing Office, Washington, D.C. 199 p.

OTA (Office of Technology Assessment), U.S. Congress. 1993b. Harmful non-indigenous species in the United States. OTA-F-565. U.S. Government Printing Office, Washington, D.C. 391 p.

Phillips, D. J. G. 1993. Developing-country aquaculture, trace chemical contaminants, and public health concerns. pp. 296–311, In: R. S. V. Pullin, H. Rosenthal, and J. L. Maclean, (Ed.). Environment and aquaculture in developing countries. ICLARM/DGTZ, Manila, Philippines.

Pillay, T. V. R. 1977. Planning aquaculture development: an introductory guide. Fishing News Books, Surrey, England. 150 p.

Pillay, T. V. R. 1990. Aquaculture: principles and practices. Blackwell Scientific, Oxford, England. 575 p.

Pillay, T. V. R. 1992. Aquaculture and the environment. John Wiley and Sons, New York. 189 p.

Pollnac, R. B., S. Peterson, and L. J. Smith. 1982. Elements in evaluating success and failure in aquaculture projects. pp. 131–144, In: L. J. Smith and S. Peterson (Ed.). Aquaculture development in less developed countries: social, economic and political problems. Westview Press, Boulder, Col.

Postel, S. 1996. Forging a sustainable water strategy. pp. 40–59, In: State of the world 1996: a Worldwatch Institute report on progress toward a sustainable society. W. W. Norton, New York. p. 40–59.

Pullin, R. S. W. 1993. An overview of environmental issues in developing-country aquaculture. pp. 1–19, In: Pullin, R. S. V., H. Rosenthal, and J. L. Maclean (Eds.). Environment and aquaculture in developing countries. ICLARM/DGTZ, Manila, Philippines.

Pullin, R. S. V., H. Rosenthal, and J. L. Maclean (Eds.). 1993. Environment and aquaculture in developing countries. ICLARM/DGTZ, Manila, Philippines. 359 p.

Rona, D. C. 1988. Environmental permits, a time-saving guide. Van Nostrand Reinhold, New York. 433 p.

Roodman, D. M. 1996. Harnessing the market for the environment. pp. 168–187, In: State of the world 1996: Worldwatch Institute report on progress toward a sustainable society. W. W. Norton, New York.

Rosenfield, A., and R. Mann. 1992. Dispersal of living organisms into aquatic ecosystems. Maryland Sea Grant College Program, College Park, Md. 471 p.

Rosenthal, H. 1994. Aquaculture and the environment. World Aquacult., 25(2): 4–11.

Roy, R. N. 1985. Consultation on social feasibility of coastal aquaculture, Madras, India, November 26–December 1, 1984. Fishery Development Series 16. NSBF/BBP, Madras, India. 125 p.

Rubino, M. C., and C. A. Wilson. 1993. Issues in aquaculture regulation. NOAA/USDA, Washington, D.C. 72 p.

Ruddle, K., 1993. The impacts of aquaculture development on socioeconomic environments in developing countries: towards a paradigm for assessment. pp. 20–41, In: R. S. V. Pullin, H. Rosenthal, and J. L. Maclean, (Eds.). Environment and aquaculture in developing countries. ICLARM/DGTZ, Manila, Philippines.

Sandy, R., 1990. After the gold rush or why aquaculture businesses fail. pp. 353–355, In: Proceedings of Aquaculture International. September 4–7, Vancouver, B.C. Heighway Ltd., Vancouver, B.C.

Shang, Y. C. 1990. Socioeconomic constraints of aquaculture in Asia. World Aquacult., 21(1): 34–35, 42–43.

Sindermann, C. J. 1986. Strategies for reduced risks from introductions of aquatic organisms: a marine perspective. Fisheries, 11: 10–15.

Sindermann, C. J. 1992. Role of the International Council for Exploration of the Sea (ICES) concerning introductions of marine organisms. pp. 367–376, In: A. Rosenfield and R. Mann (Ed.). Dispersal of living organisms into aquatic ecosystems. Maryland Sea Grant College Program, College Park, Md.

Skladany, M., and C. Bailey. 1994. Aquacultural contributions to community development in the United States. Auburn University, Auburn, Ala. (unpublished manuscript).

Smith, H. H. 1983. A citizen's guide to zoning. American Planning Associates, Planners Press, Chicago, Ill. 198 p.

Smith, I. R. 1984. Social feasibility of coastal aquaculture: packaged technology from above or participatory rural development? pp. 627–634, In: Consultation on social feasibility of coastal aquaculture. Madras, India, November 26–December 1, 1984. Fishery Development Series 16. NSBF/BBP, Madras, India.

Smith, L. J. and S. Peterson. 1982. An introduction to aquaculture development in less developed countries. pp. 1–10, In: L. J. Smith and S. Peterson (Eds.). Aquaculture development in less developed countries: social, economic and political problems. Westview Press, Boulder, Colo.

So, F. S., I. Stollman, F. Beal, and D. S. Arnold (Eds.). 1979. The practice of local government. International City Management Associates, Washington, D.C. 676 p.

So, F. S., I. Hand, and B. McDowell. 1986. The practice of state and regional planning. International City Management Association, Washington, D.C. 649 p.

Thorgaard, G. H., and S. K. Allen. 1992. Environmental impacts of inbred, hybrid and polyploid aquatic species. pp. 281–280, In: A. Rosenfield and R. Mann (Eds.). Dispersal of living organisms into aquatic ecosystems. Maryland Sea Grant College Program, College Park, Md.

Tobin, R. J. 1992. Legal and organizational considerations in the management of coastal areas. pp. 93–106, In: Proceedings of the regional workshop on coastal zone planning and management in ASEAN: lessons learned, April 28–30. ASEAN Conference Proceedings 12.

UNDP, NMDC, and FAO. 1987. Thematic evaluation of aquaculture. United Nations Development Program. Rome. 192 p.

Van Houtte, A. 1995. Fundamental techniques of environmental law and aquaculture law. pp. 451–457, In: Report on a Regional Study and Workshop on the Environmental Assessment and Management of Aquaculture Development, February 21–26, Bangkok, Thailand, TCP/RAS/2253. FAO/NACA, Bangkok, Thailand.

Verbyle, D., 1995. Satellite remote sensing of natural resources. Lewis Publishers, Boca Raton, Fla. 425 p.

Weeks, P. 1990. Aquaculture development, an anthropological perspective. World Aquacult., 21: 69–74.

Zeylmans, M. <m.zeylman@frsw.ruu.nl>. 1996. Nice Geography Servers, <http://www.frw.ruu.nl/nicegeo.html>. Revised June 11, 1996.

Postscript

JOHN E. BARDACH

The chapters in this text permit one to make a positive prognosis for aquaculture in general, but it is risky to apply it equally across the board. The tropics, inland and coastal waters, are increasingly subject to multiple-use competition, and in the temperate zone, production economics increasingly seem to permit only intensive but risky operations to thrive. In addition, there is the gamut of well over 150 cultured species, a number that appears to call for many special environmental conditions. Looked at closer, though, the important ones fall into but few species groups: the salmon and trout family (Salmonidae); the tilapias of the related genera *Oreochromis* and *Tilapia*; the Chinese and Indian carp species, all belonging to the minnow family, Cyprinidae; and the various shrimps of the family Penaeidae. Their life cycles are closed, and increasing international coordination (meetings of aquaculture societies in many parts of the world) gives hope for speedy, fuller domestication. Domestic birds and mammals have been reared for millennia, but the combination of greater information on fish life history and their physiology and behavior, (Bardach et al., 1977) coupled with advances in biotechnology, will bring aquaculture into a new mode of association between fish and humans. New species will also be added to the roster of those cultured. The quest for novelty and new delicacies by the affluent will remain, and with it the development of management techniques for new cultivars. As can be gleaned from various chapters in this book, all of this bodes well for the cultivation of aquatic animals and the opportunity to grow them at a high rate well into the next century but obstacles to this growth will be economical as much as social and environmental as technical.

The technical advances in the rearing of fish and shellfish permit improvements in efficiency and thereby also the choice of increasingly sustainable management. For instance, tilapia rearing, polyculture, and integrated aquaculture-agriculture even now produce 1g of unprocessed protein with less input in kilocalories than most other kinds of agriculture (Rawitscher and Mayer, 1977). Fewer but especially less costly inputs and savings in energy by increasing plant proteins in the rations of carnivores and the reuse of water can make for decreasing unit costs and/or higher outputs. Catfish that are carnivorous, though at the top of the food chain, can now be grown in the southern United States on a pellet diet that contains hardly any fish

Sustainable Aquaculture, Edited by John E. Bardach
ISBN 0-471-14829-6

meal anymore. These changes toward greater sustainability parallel those that we can begin to see in agriculture in several developed countries where crop rotation is increasingly practiced and high inputs of pesticides and fertilizers supporting vast monocultures are on the wane (Kloepper, 1996).

The bulk of cultured fish production comes from semi-intensive and extensive pond cultures with many small scale subsistence ponds, predominantly in Asia. These have the potential of achieving a three- to fourfold increase in output through technomanagerial improvements. But conditions for implementing these would include improved financial support and greatly improved extension services; Chapters 2 and 7 both bear this out.

Money and technical assistance need to be applied simultaneously, for applying money without understanding the social and cultural conditions, as has sometimes happened in the past (e.g., in the Pacific Islands in the 1970s and 1980s; Uwate et al., 1984), does not lead to the establishment of sound, sustainable aquacultural installations. This is now beginning to change, importantly by placing their development under the management of the most suitable agency or agencies (see also Chapter 8).

In particular, decisions may have to be made as to whether a department of fisheries or agriculture ought to have administrative oversight, as indicated in Chapter 1. Like agriculture, inland aquaculture uses land and water. Integrating aquaculture into farms is now done in various parts of Africa, as Williams discusses in Chapter 2, and it has also been successful in the river plains of southern China. Therefore, placing all farm practices under an agricultural agency seems only sensible. But, self-evident as considering this may seem, the changes necessary on local, regional, and national levels may well have to wait for the overhaul of many a bureaucracy and, importantly, for development of cadres of young, well-trained officials who understand the interplay of economics and ecology.

And now to the coast where urbanization progresses apace and space becomes increasingly scarce, both on land and in the water (Anonymous, 1996). The bulk of animals reared there are mollusks and shrimps but also predatory finfish such as groupers or yellowtail; but even the diet of filter-feeding mollusks consists of plankton that is part plant and part animal. Thus, the main economic problems in the culture of carnivores are feeds and their costs and also the reduction of waste. The latter becomes less important as one proceeds farther out to sea, but then the costs of containment increase (see Chapters 3 and 6) and with them the challenges to engineers.

On the nontechnical side, coastal aquaculture calls for solving problems of entitlement (e.g., sites in mangrove areas are usually cheap) and for resolving competition over access to sites if not also on the application of environmental controls. Disagreements abound over which several departments or agencies should have oversight, only to point to the need for integration in coastal-zone management; if a country has a Land and Natural Resources Agency, that agency is probably best suited to include a coastal unit that oversees coastal aquaculture (see Chapter 8).

In the case of shrimp, environmental control is not applied adequately. After all, the crops are grown locally for the tables of the affluent abroad, with foreign-currency returns as prime considerations; the environment and social problems, if noted at all, are left for later attention. A case in point appears to be Vietnam, where a survey report on Peneid farming in the country ends with a declaration of needs: " . . . to protect mangrove forests and prevent the eutrophication of the (coastal) water regions. The plan for management should be prepared" (Nguyen Tac An, 1995). Whether it will be followed and how it will be implemented are left unsaid. Similarly, in China, Li Sifa (1995) warns against "pollution from intensified shrimp/fish farming" and, harking back to urbanization, against "mismanaged industry and city sewage." These ill effects occurred after rapid growth of fish culture that had reached its limits. There are serious externalities that have to be regulated from a social vantage point, but as long as there is a strong demand for more fish (and shrimp) and riparian people do not (or cannot) complain about eutrophication and occasional poisonings, strict regulation will be difficult to enforce. There will be difficulty in making remedial measures such as assessing environmental carrying capacities, instituting input and output controls, applying remedies for past excesses, and the like (see Chapter 9). The chances of sustainable use of such highly populous coasts under pressure and for the aquaculture in them are not too promising.

But steps have already begun to move aquaculture farther offshore. These efforts will be faltering at first, being costly and risky (Willinsky and Champ, 1993); they will also be restricted to rearing luxury fare with an emphasis on predatory fish, not on species tied to a substrate such as shrimp. The technology is in its infancy and involves mainly problems of containing and feeding the fish. Both these exigencies have been approached from the vantage point of enclosures or of behavior modification with acoustic-feeding training, requiring fewer and less risky structures (see Chapter 3).

For containment offshore one could cite again the Israeli scheme of using modified ships to site the cages and hold the feed. Inasmuch as the life span of such structures is likely to be not more than decades, the concept but not the structure is the sustainable component here, as is true in other modes of culture (e.g., the pattern of polyculture and not its ponds). In contrast to enclosures, Japanese, Norwegian, and Swedish scientists and officials have recognized the opportunity of using behavior modification (Bardach and Magnuson, 1980) and technologies related to it as important building stones in their sea-farming strategies (NSF, 1991 and Chapter 3).

It is offshore that cultivation and the management of fish populations could merge—salmon ranching is a beginning here—leading to a broader kind of aquaculture. We have not treated ranching in this book because it would have been largely speculative to do so, but a small excursion into this subject matter is deemed appropriate here at the end.

Japan, which is after all the most fish conscious nation on earth, recognizes that it needs to engage in several activities to move fish ranching slowly and deliberately into the sea and to furnish a growing base for Japan's important fish harvest. Fisheries managers are participants in them, and the biological component is divided into a highly migratory species (tuna), a migratory species (salmon), and a static species (flounder) component, these have different applied scientific and administrative problems, but all species considered are expensive seafood items. There is attention to tight cooperation among industry, government, and applied science, perhaps more so than found anywhere else (Park 1996).

An impressive aspect of the efforts to extend aquaculture offshore in Japan is not only the connection to fisheries but that the institutions and agencies that have pooled forces recognize that the goal is far in the future and that progress in reaching it will be slow. Park (1996) also notes that only 2% of the open ocean's primary productivity is used (Pauly and Christensen, 1995) but that there are stationary patches of high fertility on the high seas. Aside from a long-term program on tuna rearing in Japan, industry and government planners are also researching a form of sea ranching open-ocean sardines that will be facilitated by remote sensing assessment of plankton blooms. Sardine larvae would be injected into the plankton blooms; however, the techniques have yet to be developed.

Such schemes, far beyond anything considered in this book, would depend on further advances in, and integration of, technologies now begun and on international cooperation exceeding that now in progress. Park (1996) advocates that the United States with its large Exclusive Economic Zone (EEZ) and high technology, should plan for a prominent role in this long-range effort to increase ocean yields.

This kind of aquaculture would be of a different order of magnitude than net pens and cages, and tied to it would be the risks and the uncertainties of large-scale manipulations in the biosphere. Yet human population increases, economic development, and increasing demand for animal protein are likely to favor measures of this kind. When properly based in science and a respect for nature, they could indeed assist in promoting an important portion of the sustainable supplies of living aquatic resources for humanity.

REFERENCES

Bardach, J. E. and J. J. Magnuson. 1980. pp. 1–31 In: Fish behavior and its use in the capture and culture of fishes. Conference Proceedings 5, International Center for Living Aquatic Resources, Manila, Philippines.

Bardach, J. E., J. Magnuson, R.C. May, and J. M. Reinhart (Eds.). 1977. Fish behavior and its use in the capture and culture of fishes. ICLARM, Manila, Philippines.

Kloepper, J. W. 1996. Most specificity in microbe-microbe interactions. BioScience, 46(6): 406–409.

Li Sifa. 1995. Opportunity and crisis of sustainable development of aquaculture in China. p. 243 In: Proceedings of Sustainable Aquaculture '95, PACON '95. Pacific Congress on Marine Science and Technology. June 11–14, 1995, Honolulu, Hawaii.

Nguyen Tac An. 1995. The development and management of the peneid farming in Vietnam. p. 253 In: Proceedings of Sustainable Aquaculture '95, PACON '95. Pacific Congress on Marine Science and Technology. June 11–14, 1995, Honolulu, Hawaii.

Park, P. K. 1996. Fisheries technology cooperation in the twenty-first century. pp. 539–551 In: Proceedings U.S. Japan Natural Resources Panel, U.S. Department of Commerce, Washington, D.C.

Pauly, D. and V. Christensen. 1995. Primary production required to sustain global fisheries. Nature, 374:255–258.

Rawitscher, M., and J. Mayer. 1977. Nutritional outputs and energy inputs in seafoods. Science, 198:261–264.

Uwate, K. Roger. 1990. Market, technical, financial, economic and organizational considerations for preparing aquaculture feasibility studies in the Pacific Islands region. Pacific Internatioanl for High Technology Research, Honolulu, Hawaii. 232 p.

Willinsky, M. D. and M. A. Champ. 1993. Offshore fish farming: reversing the oceanic dustbowl. Sea Technology, single issue. 4 p.

The World Resources Institute, the United Nations Environmental Program, the United Nations Development Program, the World Bank. 1996. World Resources, a guide to the global environment: The urban environment. Oxford University Press, New York/Oxford. 364 p.

INDEX

CSCTCA 639
.8
S964